The Beginning of the Path

to Human Extinction, and HOW TO GET OFF IT -

Notes on a Paradigm Shift

by

Scott C. Haley

Author's professional association memberships: Society for Conservation Biology; International Society for Ecological Economics; American Association of Geographers; International Society for Environmental Ethics.

DEDICATION

Kudos to Paul R. Ehrlich, ecologist, Professor Emeritus at Stanford University, and still semi-active in the Stanford Center for Conservation Biology (as of March, 2022). Kudos, too, to all science publishers who allow "Open Access" to their crucial scientific information. Hats off as well to: John Muir, Aldo Leopold, Roger Revelle (who studied anthropogenic climate disruption in the 1950's), Rachel Carson, David Brower, Barry Commoner, E. O. Wilson, William E. Rees, Anne Ehrlich, Herman Daly, Ira Judd (decades ago at Arizona State U.), Virgil Baker (also formerly at ASU), Diana Beresford-Kroeger, Brian Czech, Joshua Farley, Robert Costanza, Kenneth Boulding, Bucky Fuller, Richard Heinberg, Suzanne Simard, Lester R. Brown, John Cairns, Jr., and many others for inspiring me to pursue an understanding of ecology, particularly human ecology. During the last ten to fifty years, without them knowing it, their works have been crucial to my quest for ecological knowledge. I dedicate this book to them. My deepest thanks to all.

Important Note to Readers

As you can see, this book is not a long magnum opus. Its purpose is to show the reader the urgency and seriousness of the most multi-faceted, dangerous crisis our world ever has faced. *More importantly,* it summarizes **solutions** to that crisis, and ways we all can help implement many of those solutions.

The text you're reading is not an academic thesis or scientific journal article. It's written for both lay people and those who have varying degrees of natural science and social science expertise. Despite that and its relatively short length, be assured that scientific rigor was not sacrificed during its creation. Every effort has been made to make the book both accurate and up to date. This work is not original scholarly research by me; instead, it's the communicating and synthesis of same done by others. [In that regard, it's somewhat similar to what Rachel Carson did in her piece, *Silent Spring*.] You will, however, find original *ideas* of mine here; but such do not constitute empirical research.

For background and context in writing this book, I relied on my own observations and experiences during years of both office and field work in various aspects of ecology, environmental regulatory compliance, and environmental geography.   The core of "library" research for the book, though, is from a review/study of well over 250 scholarly journal articles on appropriate subjects. Then, too, numerous textbooks, professional conference reports, scientific news reports, and relevant essays were utilized. See the "References" section for key sources.

Rather than spending at least five years writing a 1200-page comprehensive text on the subject, I chose to go this route. Why? Time is of the essence. Consequently, don't expect to find every single aspect of the socio-eco-econ-ethical Crisis addressed here. I've covered what I believe are key points. The crisis is not only in the future. **We're in it now.** Tick-tock!

It isn't only climate disruption that's a monumental problem, or biodiversity loss, or soil degradation, or toxics in our environment, or the overharvesting of fisheries and forests. Nor is it only the loss of kelp forests and coral, or *viral disease spread*, or acidification of the ocean, or the worldwide debt bubble, or the lack of job security, or the Rat Race of the pursuit of infinite growth on a finite planet, or social inequality, or an economic playing field that isn't level, or racism. Nor is it only the decades-long problem of nuclear waste storage, or the threat of nuclear weapons, or ongoing perpetual war. Nor is it only overconsumption of goods, or overpopulation, or food insecurity, or societal disintegration, or habitat destruction, or the lack of ethics in uncounted situations. Finally, nor is it only the Edward Bernays style propaganda coming from the State-Corporate-Military-Financial Complex. Unfortunately for us, *it's all the above...and more.*

There is a way to get off that path. We can have a sustainable future in which life on Earth not only survives, but thrives as well. Getting there will be a tough row to hoe. That's mostly because, in general, we humans – particularly in developed countries - have been indoctrinated to believe numerous false concepts. To get where we need to be will require a paradigm shift in ethics, values, and our worldview in general.

Everyone is encouraged to read through the "References" section, and track down any sources of particular interest. Some truly amazing generation of ideas and research is taking place now, and has been going on in the fairly recent past. In addition, because history is important in the study of any subject, some references go back decades ago. Listing all the sources used would have been cumbersome to say the least, and frankly, unnecessary for this type of book. I chose what I thought were the best of the bunch, about 100 or more.

Happy Trails,

Scott C. Haley

Table of Contents

A Bit of Background

Nuclear Weapons

Nuclear Wastes

ICAN

The Big Picture Problem

The Biofuels Problem

The Wind Turbine Problem

The Solar Energy Problem

Expansion of the Human Footprint

Loss of Natural Habitat

The Dilemma

Preface and Principles

Conflict

The Steady State Economy

Awareness and You

Human Population

Worldview

Either Mother Nature or Humans Will Solve the Problem

Is Human Extinction Possible?

The Core of Ecoethics

Micro and Macro Levels

Small is Beautiful

Universal Property

Carbon and Climate

Biodiversity

Our Toxic Environment

Nuclear Idiocy

The Plastic Problem

The Social Issues

Energy

Regenerative Agriculture

Fresh Water Crisis

The Food Crisis

Ethics

Politics and Sustainability

Our Greatest Challenge

Synthesis

Sustainability

Chapter 1: Introduction

Preface

This is a story involving you, me, everyone else in the world, all nonhuman life, our environment, science, and ecoethics. As a species, we are beset by a Crisis which was decades in the making. We don't have decades to correct the unprecedented mess we've made for ourselves. Plus, some of the negative effects of that Crisis often are not readily visible. With a lot of dedication and a bit of good luck, though, we'll pull through. Or, perhaps better stated: we'll survive *and thrive* **if** we undergo a paradigm shift in values and actions. The need for and the specifics of that shift are the subject of this book. Solutions to our Socio-Eco-Econ-Ethical Crisis are known and/or are being researched. This text will show you how to help implement some of those solutions.

We shall not spend too much time detailing all the horrendous problems currently being faced by all life on Earth. Many experts in their respective fields already have done that. We'll do it to some degree where necessary, but mostly the concentration here will be on mitigation of and/or solutions to those problems. One thing is clear: the various problems are all systemic, and thus interconnected.

It's not a question of saving the Earth. Our planet will survive. It's a matter of *organized* human existence surviving and thriving. It's time for all of us to overcome ignorance, apathy or hopelessness, Edward Bernays style propaganda, and an ethics based on materialism and consumerism. Our world is no longer "full" of natural resources, nor is it any longer relatively "empty" of human beings. Those days are long gone. Ask Herman E. Daly about that. More on him later.

My first job as an environmental geographer was as one of the main field researchers and authors of the "Water Pollution Control and Abatement Plan for Drainage Basin 15, State of Washington". That position was at a landscape architecture consulting firm in 1973 in Seattle, Washington. The company had a contract with the Puget Sound Governmental Conference to produce the Plan.

That experience and various later positions in teaching, consulting, and a period of almost seven years with the Sacramento County Environmental Management Department led me to the following conclusions.

1. Despite some forward progress, Humanity was (and is still) heading for more and more ecological disasters. Accompanying them would be (and are) more social, economic, and ethical problems.

2. Some of the reasons why include materialism, unlimited consumption, unlimited growth, corporatism, neoliberal policies, and politics in general.

3. The main reason, though, was/is **a lack of ethics... specifically, ecoethics.**

Our environmental/ecological/social problem on Earth is not only material in nature, it's metaphysical, spiritual. Spiritual not in the sense of organized religion, but rather in the sense of ecoethics, life purpose, relationships to nonhuman life, and values which benefit/respect all biotic & abiotic parts of this amazing planet. It's a question of where and how to direct our energy during our short time in this physical, cosmic dimension.

Ecoethics

Philosophy is the study of knowledge, being, and reality. One of its branches is Ethics, the study of conduct and behavior. Ethics is concerned with "right" and "wrong", and in particular, with what is "good" for individuals *and* society. Its three primary principles are: respect for persons, beneficence, and justice. A simple definition would be: a code of conduct for human good. Socrates was the Father of Ethics in the western world.

It has taken about forty to fifty years for neoliberalism and corporate globalization to bring us to what might be the beginning of the path to **human extinction**. *If not that*, it is at the very least, the path to the end of *organized* human existence. In other words, the beginning of the end of civil society. Neoliberals (and many others) adhere to ethics which are anthropocentric, not ecocentric. Nonhuman life too often is treated as not very significant, mostly disposable, and certainly not spiritual in any sense of the word. As a result of such ethics, nonhuman life which is crucial to our survival is in decline.

The Powers-That-Be in most countries around the world have opted (for decades) to implement neoliberal economics in their Lands. [See Appendix I] The result has been severe, worldwide environmental and social damage. Despite efforts to mitigate that damage, it continues to this day. Most ecologists agree that's because rather than Aldo Leopold's "Land Ethic" (or something similar), countries have chosen to follow the anthropocentric, neoliberal path of unlimited growth and overconsumption. In poorer Lands in general, the path appears to be one of emulating the richer countries.

Even in countries that are attempting to "go green", the prevailing ethic often is "growthism" not sustainability. More and more perpetual growth apparently is mistakenly viewed as "good" for society. History disagrees with that assessment.

Despite the shrinking of *nonrenewable* natural resources and the overharvesting of *renewable* natural resources, human populations (in general) have been indoctrinated to accept without question the idea that economic growth must continually expand for the good of individuals and society. "Going green" supposedly will solve the problems of ecological/environmental damage. That most likely will help a bit, but only a bit. Much more is needed.

Here's why.

1. The Law of Conservation of Matter & Energy states that *energy and matter cannot be created or destroyed*; they only can be transformed from one type to another.

2. Because of entropy, every time energy is transformed some of it is lost as dispersed heat. "Lost" means it's no longer available to do "work" because it's too dispersed.

3. *Materials cannot be 100% recycled.* While mass is conserved, micro amounts are dispersed in the environment during recycling. Lost, so to speak. So, nonrenewable mineral resources (e.g., rare earth minerals which are gobbled up by High Tech industries, and oil) will continue to shrink. [Some researchers claim it is theoretically possible to recycle all materials 100% because Earth has a constant inflow of solar energy, and thus is not a closed system. They seemingly are ignoring or missing a few facts: Earth *is* a closed system in terms of *material/mass*; also, currently, the technology for complete recycling of all materials does not exist; plus, any method now imaginable for such recycling would cost too much in both energy and finances to be feasible; and, both the scale (in terms of amounts needed to be recycled) and the time involved –

never mind the cost – make the proposition implausible.]

4. Although solar energy is "renewable" (super-abundant might be more accurate), it obviously has to be captured. If the demand for solar power exceeds the rate of the in-flow *and capture* of solar energy, then sooner or later, consumption has to be limited.

5. While the stock of solar energy (in the sun) is super-abundant, the in-flow and capture of it on Earth are limited. We don't get any useful amount of it at night or on densely cloudy days. *On any day, we don't capture most of it.*

6. Whether "green" or not, all material economic growth consumes resources and produces wastes. That's even true of service industries such as health care, education, taxi service, tourism, the financial sector, and others.

7. Ever-increasing growth results in an ever-increasing flow of "throughput": resources from Earth's ecosystems go through the economic subsystem, and then out to natural ecosystems as wastes. If "growth" never ceases, sooner or later throughput will exceed Earth's resource regeneration and waste assimilation capacities, as well as exceed recycling and reuse by humans. That's called ecological "overshoot", and we're already there. Many ecologists agree: our activities have exceeded Earth's carrying capacity. A few studies have postulated that our planet can comfortably accommodate about three billion humans and all our activities.

8. Renewable resources, such as ocean fish and trees, are being overharvested already. More and more economic growth will exacerbate the problem.

Some analysts believe that because this is the "Information Age", soon economies won't need to rely on natural resources for commerce much at all. They seem to forget that most of the "information" being utilized and sold/traded deals with physical, natural resources – lumber, fuels, fiber, water, grain, livestock, metals, granite, and the like. Plus, we can't live in houses built only with information, or wear jeans made of information, or eat information.

All the above means sustainability presently is a fairly long way off. It also means that, first and foremost, we have a massive ethical problem. Without a paradigm shift in the cultural and ethical thinking & behavior of humans, serious ecological damage on Earth will continue. Why? Because the Powers-That-Be are firmly locked onto the path of unlimited growth and overconsumption. The only way to change that is *from the ground up, not the top down*... and, in my opinion, that won't happen until a majority of us (or at least a significantly large number of us) adopt what ecologists call ecoethics. Only then will people gravitate to peacefully demanding the implementation of ecological economics, or something similar. The good news is: that already has started to take place in large parts of Europe, in a few regions in South America, and also in a few other places scattered around the globe. After over fifty years of observing the USA in this regard, in my opinion we are lagging behind...way behind.

Ecoethics is ethics with an ecocentric - rather than anthropocentric – worldview. *It's a holistic view*. That means natural ecosystems are viewed as having **intrinsic** value, and as all being interdependent. They are not viewed merely as areas from which natural resources are to be extracted for human use and into which wastes are deposited. In short, this all means when we must extract resources, we should do it with great care and minimal damage. When we visit or live in natural areas, we should take care to not unnecessarily cause harm to the biotic and abiotic elements of Nature. It also means we have to re-think the concept of unlimited growth in a finite habitat. *We must view Nature as critical to our survival*, not only something nice to enjoy every so often. Currently, that's not the prevailing view of most people. At least, it certainly seems so to me. The Eco-Crisis we're now in appears to confirm that opinion.

Unlimited Growth and Overconsumption

It should be axiomatic that the concept of perpetual, unlimited economic growth plus more and more consumption on a finite planet is delusional thinking. Instead, that concept appears to be the unquestionable mantra of mainstream/neoliberal economists and other policy makers around the world. I believe it came to be that way because it started when our world was relatively empty of humans, and *relatively full* of natural resources.

In order to illustrate the above point, let's consider the fishing industry. It's an important economic activity, and crucial in terms of helping to feed the world. Decades ago, proponents of unlimited economic growth apparently believed the only limitations to the industry were capital and technology. After all, fish are a renewable natural resource. Only more fishing boats and better technology were necessary to keep expanding commercial fishing. After years and years of such expansion, plus more and more pollution finding its way into bodies of water, today we have two results: severe overharvesting; and significant reduction of fish populations. Building more boats with better fish-capturing technology will not solve the problem. That is the probem.

When I was a freshman at Colorado State U. (**1961**), the global human population was only three billion. Perceptions of future, unlimited possibilities were quite bright then. Even though it was fairly pervasive, widespread environmental degradation largely was unrecognized by most people. Natural resources seemed to be super-abundant, and in a perpetually unlimited supply. Overall, the biophysical world appeared to be almost limitless. The idol of unlimited economic expansion was "worshipped" around the world.

In **1968**, a book by Paul R. Ehrlich & his wife, Anne, *The Population Bomb*, was published. It was a best-seller, and made the points that this planet is finite, the natural environment will be degraded even more significantly, and the availability of resources will not keep pace with overpopulation. The book was attacked rather viciously by believers in unlimited growth.

In **1972**, after a two-year study by M.I.T. researchers (utilizing a large, mainframe computer), the results were published in a book titled, *The Limits of Growth*. Like the Ehrlichs' book, it was a best-seller. And it was attacked even more fiercely. There's a fascinating story about it, and the author, Christopher Ketcham, is one helluva writer. See the details here---

https://psmag.com/magazine/fallacy-of-endless-growth .

I recently watched a 2011 British documentary, **"Consumed - inside the belly of the beast"**. It effectively illustrated humanity's cultural evolution to the stage at which we find ourselves now: lost in materialism, consumerism, short-term shallow thinking, the pursuit of unnecessary prestige, and trying our best to ignore the destruction of the ecosphere. It also put forth the proposition that this stage is a temporary glitch in the development of the species, *Homo sapiens*. The makers of the film see a future shift to sustainability and ecoethics. Let's all hope that's the case for us.

The film detailed how, over decades, we've been molded by advertising and propaganda to believe that consumer goods can bring us meaning, prestige, contentment, fulfillment, and the big kahuna, happiness. The key is to buy more and more goods. Over the years, as we've come to realize ultimately none of that is true, our discontent, anxiety, and emptiness all have increased. During the same time, we've been trained (in a sense) to seek instant gratification, all the latest tech gizmos (to be replaced every year or two), and to desire having "the latest thing". Now, we're at the point of a line from an old Rolling Stones' song: "I can't get no satisfaction.". On top of all that, too many people seem to believe that Nature is nice, but not especially important... and not really necessary.

In recent years, both mainstream economists and corporate America have used all the above to double-down on their promotion of perpetual, unlimited growth. Any problems regarding natural resource depletion or ecosphere damage, they say, can be handled by new technology and/or the substitution of one resource for another. Some mega corporations even have advertised their new "green" initiatives concerning corporate operations. There's one big problem with all these solutions: they are all within the framework of **continuing unlimited economic growth and consumption**... on a finite planet.

Ever-increasing economic growth means ever-increasing throughput. [Again, "throughput" is the total flow of resources from the Earth ecosystem to the economic subsystem... *and then back to the ecosystem as waste*.] One doesn't have to be a genius to understand that more & more & more of such a system is **unsustainable**. It's folly to believe that undiscovered, new technology and/or substitutable resources will prevent the collapse of our natural life support system. *That collapse already has started*. Much more than we've been trying to, we need to mitigate it now. We must undergo a major shift in ethical and cultural values. The old paradigm is killing us... literally... both directly and indirectly. It's time to implement steady state, ecological economics.

Ignorance by Economists of Biophysical Constraints

It appears to me that most mainstream economists have had little to no higher education in the natural sciences. In the essay at the link above, Ketcham quotes a number of them, as follows.

1. Oxford U. economist, W. Beckerman: "[There's] no reason to suppose that economic growth cannot continue for another 2,500 years.".

2. Harvard economist, C. Kaysen: "[Some studies show] the Earth's available matter and energy could support a population of **3.5 trillion**...".

3. J. Simon (deceased), University of Illinois economist, stated in 1992: "We now have... the technology to feed, clothe, and supply energy to an ever-growing population for the next **7 billion** years.".

Those beliefs are, of course, sheer nonsense. I can surmise only that they are due to an almost total ignorance of natural science.

In contrast, ecological economics fully recognizes biophysical constraints and the negative impacts on our natural life support systems of pursuing unlimited growth. From *An Introduction to Ecological Economics (1997),* by Robert Costanza, Herman Daly, et al.: "The basic problems...include: ...highly entropy-increasing technologies that deplete the earth of its resources and whose unassimilated wastes poison the air, water, and land... ". [Fair use quote]

Equity

It's all well and good to say, in the free, developed-world countries we have a democratic system which ensures everyone's chance to pursue happiness and fulfillment; however, the reality is as follows.

1. The playing field too often is not level; it's not even close to level.

2. Resources (especially financial resources) often are not allocated **fairly**.

3. Many countries (including the USA) don't appear to understand the value of having all citizens educated to the highest degree their capabilities and desires allow. Some other, more enlightened nations provide access to universal higher education at (for the most part) no cost to the student.

4. Adequate health care for many people (including many in the USA) is not available, or is too expensive.

5. Neoliberal politics and corporatist policies (both public and private) greatly favor the Upper Class. The excuse given is that the Rich supply jobs to everyone else. Wealth supposedly "**trickles down**". That's more often not true than it's true. Even when it does happen, the jobs too often are temporary and/or part-time, and/or low-paying.

6. Too many poor people often are relegated to living in neighborhoods which are much too close to the "sinks" of economic throughput **wastes**. The resulting exposure to air pollution, water pollution, and land/soil pollution negatively affects their physical

health, mental health, and general well-being.

7. Perpetual wars and insanely bloated defense budgets siphon limited funds away from infrastructure repair/replacement and from **social safety nets**. Primarily, they benefit Mega Banks and other Mega Corporations. The wars often are the result of **shrinking natural resources**. Stronger countries want guaranteed access to them.

In the financial sphere, the central banks and about thirty mega banks around the globe have created a monetary system which is so convoluted and complex that *accountability is difficult to locate*. On top of that fiasco, the big players are so interconnected that the entire global financial system is extremely fragile. Any one of hundreds of unexpected, unplanned events (a "black swan" event) can bring the whole thing crashing down. Those who suffer the most are on the bottom three-fourths or so of the economic ladder. The Super-Rich emerge from any such disaster relatively unscathed.

Conclusion

Pursuing unlimited economic growth, overconsumption, a lack of equity, perpetual wars, essentially unlimited population growth, and neoliberal policies in general have resulted in:

1. an ecological crisis probably never before imagined (shrinking natural resources, damaged or destroyed ecosystems, disease proliferation, an excess of greenhouse gases, increasing pollution in general, a loss of necessary biodiversity, etc.);

2. gross social and income inequality (due to a lack of equity and ethics);

3. a growing discontent with both public and private institutions;

4. a global market which – according to Professor Guy Standing's many years of research – has created an emerging new class of workers, the *Precariat*. These people now lead lives which are extremely precarious due to constant economic insecurity. [See the end of "Appendix II", and "References".]

5. an increasing sense of despair over the condition of humanity.

It's time for a significant change in present-day economics, equity, and ethics. Ecological economics is one part of a sustainable and much needed path forward. Later on in this book, we'll explain how eco-econ literally can help save the world, so to speak. In reality, we needn't worry about the "world" (meaning, our planet). It will survive. As I said earlier, what we must save is *organized human existence* on the planet. No matter what your political and religious affiliations are, I'm sure we all can agree that no one wants to see society devolve into a *Mad Max* scenario. To avoid it, we must be concerned for not only us, but nonhuman life and the abiotic environment as well.

[NOTE: the core of ecoethics and how it relates to you are discussed in Chapter 10.]

Off and on, I've been studying various aspects of this predicament for about fifty years (starting in 1972 at Arizona State U.), so please believe this: humans are facing the greatest Crisis in our history, and climate disruption is only a part of it. The "experts" think we have until 2050 to mitigate the problem & thus avoid total societal break-down. In all probability, I believe that benchmark should be moved to the year **2035**. Perhaps earlier. Why? Because much more than climate is involved.

It's time for all of us to take this seriously. Much can be done on the micro scale, and much "lobbying" (of sorts) needs to be done on the macro scale. Politicians will engage in window dressing efforts *unless there's an outcry from the ground up.* [See "Appendix III".] Both public and private, the Powers-That-Be in the Upper Crust will do little to nothing unless Main Street makes it abundantly clear that things must change. In general, history confirms that contention.

Chapter 2: Biodiversity

Importance to Ecosystems

Understanding biodiversity requires at least a rudimentary understanding of ecosystems and an appreciation of their complexity. An ecosystem is a <u>dynamic</u> system made up of the physical environment and the interacting organisms which occupy it. Relative to humans, key factors involved are as follows.

Ecosystems are always adjusting, always striving for a type of homeostasis (balance). When we humans interfere with the functioning of them, we should do so with the least damage possible under the particular circumstances. Why? Because: 1) they all, *directly or indirectly*, contribute to our health, prosperity, and well-being; and 2) human existence is not possible without the functioning of natural cycles, and those cycles are found in ecosystems. Even if you're in artificial surroundings all day and night, you still need clean air, clean water, decent food, shelter, protection from floods, disease, etc. That all is provided by properly functioning ecosystems. "Everything is connected to everything else." [Barry Commoner's first law of ecology in his classic book, *The Closing Circle*. See "References".]

NOTE: In the mid-1970's, I used *The Closing Circle* as the textbook for an Advanced Ecology course I taught at South Kitsap High School (one of the top high schools in the State of Washington at that time). I found it then, **and still do today**, to be one of the most informative books ever written on Human Ecology in particular, and Ecology in general.

It was written for lay people, and in my opinion, should be required reading for every student everywhere.

Professor Commoner's Four Laws of Ecology---
1. Everything is connected to everything else. [More on this later.]
2. Everything must go somewhere. [Wastes don't simply disappear.]
3. Nature knows best. [More on this later.]
4. There is no such thing as a free lunch. [Example: there's a price to pay when we overload the atmosphere – which is treated as a "free" waste sink - with pollutants.]

Ecology is the branch of Biology which studies the relationships of organisms to one another and to their physical surroundings. **It's essentially the study of ecosystems.** There are many sub-branches of Ecology, e.g., wetlands ecology, grasslands ecology, desert ecology, human ecology, mountain ecology, forest ecology, etc.

As the name implies, **biodiversity** simply is all of the various interconnected life forms in an area – from a small ecosystem on up to the largest ecosystem, Spaceship Earth. The greater the biodiversity, the more resilient and stronger the ecosystem. Why? First, because in order to be resilient and sustaining, food webs must be varied and complex. If one part of the web gets damaged or wiped out by plant/animal disease, weather, parasites, habitat loss, or other factors, then other parts of the web often can fill the gap. Secondly, the greater the variety of plants in a terrestrial ecosystem, the more shelter from the elements animals will have... and thus, the greater the variety of animals to support the food web. Even in aquatic ecosystems, plant variety plays an important role in the resilience of the system. Thirdly, the greater the variety of life in any ecosystem, the more complete the roster of nutrients available for plant life (the primary producers) will be when the old life dies and decays. A loss in biodiversity means the ecosystem will be damaged to one degree or another. The greater the loss, the more the damage.

In any ecosystem, food webs (sometimes called food chains) have different trophic levels. A trophic level is a group of organisms which occupy the same level in the web. Classifications vary a bit, but essentially there are five trophic levels:

1. primary producers – green plants and algae – the organisms which can capture the sun's energy during photosynthesis and store it as glucose (the fuel of most all living things);

2. primary consumers – herbivores – the organisms which eat plants only;

3. secondary consumers – omnivores & carnivores – which mostly eat herbivores;

4. tertiary consumers – omnivores & carnivores – which primarily eat the secondary consumers; and

5. apex predators - have no natural predators, and are at the top of the web/chain. The top predators eat just about anything they so desire.

The entire arrangement is known as the Energy Pyramid. Along with incoming solar energy, the primary producers at the base of the pyramid and a fertile, well-structured soil obviously are crucial. At each higher trophic level after that, only approximately **10% of the energy consumed– stored in "food" – is converted into biomass** (physical body parts). The rest is lost/dissipated as heat, or used for movement, bio-electrical functions, or other functions of the body. At each level, decomposers (e.g., bacteria, fungi) are working to keep the nutrient cycle going. This all contributes to a functioning and healthy ecosystem. Biodiversity loss reduces the health of an ecosystem.

So, why is that important? Humans depend upon "ecosystem services" for our survival. Some of those services are as follows: the providing of food, **fresh water**, fiber, metals, medicines, and timber; the formation of soils and soil fertility; flood control; **oxygen** from green plants via photosynthesis; the prevention of soil erosion; the **prevention of disease outbreaks**; pollination; biogeochemical cycling; and more. Keeping damage to ecosystems as minimal as possible is critical to the continuation of organized human existence on Earth. Decreases in biodiversity do not bode well for the future health of ecosystems. We must wake up people to that fact.

Biodiversity Loss

To the best of our knowledge, there have been five mass extinctions in Earth's history prior to the one we're in now. None of those five were caused by humans. The sixth is being caused by us. How do we know we're in a mass extinction event? Consider the following.

1. According to the World Wildlife Fund's *Living Planet Report 2020*, populations of vertebrates have dropped in numbers (on average worldwide) by **68%** since 1970. That means various species are declining in a relatively rapid manner, and are well on their way to extinction.

2. The same report states: populations of vertebrates in freshwater habitats have decreased in numbers by an average of **84%** since 1970.

3. A 2019 article in the journal *Science* reports that bird populations in North America have decreased in net numbers by three billion since the 1970's.

4. In a 2020 news article by the Natural History Museum, it was reported that the Royal Botanic Gardens conducted a study which found that about **40%** of plants (worldwide) are on the brink of extinction.

The driving forces (or "drivers") of the above events are fairly well known to researchers. They are: habitat loss, especially in poorer countries where vast amounts of wildlands are converted to farmlands or urban spaces; poaching; overharvesting; toxic pollution; disease; introduction of exotic species (which compete with and overcome the native species); monoculture forests*; desertification, and climate disruption. Either directly or indirectly, humans are responsible for all the drivers of this Sixth Mass Extinction.

*When thinking about biodiversity loss, imagine a natural forest. It has a rich variety of trees, shrubs, grasses, mosses, lichens, fungi, animals, and microbes. It's a highly complex ecosystem which provides many "services", including those well-known, such as the prevention of soil erosion, the production of oxygen, and the trapping of the greenhouse gas, carbon dioxide. Because of its diversity, a natural forest is resilient. It can bounce back from the onslaught of many destructive factors.

Now imagine that a timber company clears it. It's logged off, and most of the remaining vegetation (shrubs, etc.) is burned or otherwise removed. Then, in place of the previous forest, trees are re-planted on what essentially is now a timber company "plantation". Usually, only one or two species are grown. Undergrowth (mostly shrubs) is kept to a minimum, or eliminated completely. In other words, biodiversity has been greatly reduced. In addition, pesticides often are used. Some portion of them find their way into the soil, where they negatively impact any number of necessary soil microbes and fungi. This is a bare bones description of how "forest monoculture" operates.

Which of the two forests – the original, or the monoculture "plantation" – do you think will be more resilient? Which one will shelter and nourish a wider variety of organisms? Which will contribute more to soil fertility? [Artificial fertilizers generally are not good for soil microbes or soil structure. They, along with herbicides and other pesticides, commonly are used on tree plantations.] Which one will better prevent soil erosion? Which one will better resist tree-boring or leaf-eating beetles (which are usually specific to each species of tree)? You get the idea.

Consider, too, the following. Humans cut down (worldwide) an estimated **fifteen billion trees** per year, and re-plant only **five billion**. About half the forests which once covered almost half Earth's land surface are now gone. Only approximately one-fifth of our planet's old growth forests remain undisturbed and in pristine condition. We simply are not sustainably managing a crucial natural resource.

Forests have been called the "lungs" of the world. Luckily for us, they work differently than do animal lungs. Trees take in CO_2 and "breathe" out O_2. It's usually easy to take adequate oxygen for granted, but we really shouldn't. All green plants should be fully appreciated and valued as absolutely necessary to life on Earth. [A possible exception to that would be noxious weeds, such as various thistles & mesquites – both of which are injurious – and poisonous/deadly nightshade.] Arguably, trees should be at the top of the list of green plants to be revered. If for no other other reason, because they provide us with oxygen and are necessary carbon "sinks" (and that's critical to the climate problem). It has been estimated that **deforestation** accounts for approximately 20% of the increase in atmospheric carbon dioxide.

At times, we no longer seem to fully appreciate that green plants in general truly are a miracle of Nature. As you may recall from your high school Biology class, the process of photosynthesis is one of the main reasons life is possible on this planet of ours. In the presence of sunlight and chlorophyll, plants convert carbon dioxide and water into glucose (food) and oxygen. The balanced chemical equation is: $6CO_2 + 6H_2O +$ (energy & chlorophyll) $\rightarrow C_6H_{12}O_6 + 6O_2$. The value of that process is incalculable. Photosynthesis is a colossal accomplishment of evolution.

Soil

Except with those humans involved in agriculture, soil is perhaps the least understood and appreciated of all our natural resources. Most folks refer to it as "dirt", and don't give it much thought beyond that. People in some parts of the petrochemical industry appear to have no regard for its complexity and value. I base that statement on the products they develop and distribute for use on plants and in the soil – primarily artificial fertilizers, insecticides, and herbicides (petrochemicals).

Along with sunlight, green plants, oxygen, and water, soil is crucial to all life on Earth. It's the foundation of all food webs, including those which are aquatic. It's the mother of all other natural ecosystems. Even the oceans depend to a degree upon life-sustaining minerals and other nutrients from soil/land runoff.

A good loam topsoil has the following composition: 45% minerals, 5% organic matter, 25% air, and 25% water. In that environment is found a *richly diverse*, highly complex community of life forms. There are the well-known forms such as bacteria, fungi, various insects, mites, worms, moles, etc. Then there are those not so well known: nitrogen-fixing bacteria, actinomycetes, nematodes, mycorrhizal fungi, and protozoa, just to name a few. All of them make up a **synergistic & symbiotic system** which nourishes not only each other, but also above-ground plants. NOTE – Mycorrhizal fungi are especially important. "Mycorrhizal" refers to a plant's root system. These fungi live on the roots of plants in a mostly symbiotic manner. Their hyphae – fungal feeding tubes – allow the host plant's roots to take in more water and nutrients than they otherwise could. The practices of *industrial* farming – with their use of synthetic fertilizers and herbicides – are destroying these valuable fungi. Thus, crop yields ultimately are reduced.

The mineral portion of soil is composed of various combinations of sand, silt, and clay. Individual clay particles are so tiny they can be seen only with a high-powered microscope. As such, they exist in what's known as a colloid, or in colloidal form. In that state, they have a net negative electrical charge. This means they attract positively charged mineral nutrients such as calcium, magnesium, iron, potassium, etc. The nutrients thus are prevented from leaching down into the lower depths of the subsoil, out of reach of many plant roots.

Humus – partially decayed organic matter – also exists in colloidal form. It, too, holds nutrients in the topsoil where they can be absorbed by plant roots. The colloids making up both inorganic clay and organic humus are the most chemically active parts of any soil. They are crucial to soil fertility. Humus is a great source of nutrition for plants, and by releasing the nutrients *slowly* during decomposition, the plant roots are not "burned" when the nutrients are absorbed.

Soil structure – the way in which individual soil particles bind together to form aggregates – is important for two main reasons. It affects the number and sizes of pores, crevices, and cracks below the surface, which affect the number and variety of life forms able to live there. Plus, structure affects the amount of air and water found in soil. The second primary reason for its importance is that it affects the amount of soil erosion which can occur when there's too much water. Soil particle aggregates (known as "peds") can be blocky, plate-like, cylindrical columns, or granular. Too much water in the soil basically can result in the total collapse of structure. Erosion follows.

Problems arise when soils are polluted (killing essential microbes and fungi in particular), overly compacted, eroded, stripped bare of plant cover, and subjected to any other type of degradation. In spite of our learning some lessons during the Dust Bowl days of the 1930's, our soils have been degraded fairly recently in too many places all around the world. That has contributed to a reduction in biodiversity both below and above ground.

Some of you may be aware of the "Green Revolution" of the late 1960's and the 1970's. It was so called not because it was ecologically Green, but rather because it resulted in a tremendous increase in food production by the first trophic level of the human food web – cereal grains, such as wheat, rye, barley, rice, oats, etc. That increase was due to three factors:

1. development of high-yielding crop strains by agricultural researchers;

2. development and field application of artificial fertilizers and pesticides by the petro-chemical industry and farmers; and

3. massive use of field irrigation.

In addition to increased food production, something else happened with the Green Revolution. *It wasn't good for life on Earth*. With the overuse of artificial fertilizers and pesticides (petrochemicals), so began the slow degradation of soils. Diverse soil microbe populations were reduced, and those populations are critical components of both the soil ecosystem and above-ground ecosystems. Soil structure was damaged, making soils more subject to wind and water erosion. Decades of all this have resulted in an unintended consequence, the degradation of soils worldwide. Also, the groundwater aquifers of the world were seriously depleted. For the most part, groundwater is not stored in soil. Instead, the reserves are in the pores and crevices of solid rock. It takes hundreds of years to fully replenish those aquifers, and about 40% of global irrigation is from groundwater. In the midwestern USA, the Ogallala aquifer is predicted to be 70%

depleted by 2050. Farmers in eight States (from South Dakota to Texas) use 90% of the Ogallala water pumped.

In the Big Picture, Key Precursors to Solutions

The bottom line is this: rapidly increasing biodiversity loss is an unfolding catastrophe. While efforts are ongoing to stem it in various locations all over the world, we as a species simply are not doing enough in that regard. Why? I believe the root cause is **a lack of ecoethics** in too many economists, politicians, super-rich billionaires, big corporations, factory/industrial farmers, and people in the Financial Sector. One might say: those who are the Movers & Shakers of the world. Too many of them act as if the economy and finance are the be-all and end-all of human existence. They seem to have little regard for the value & complexity of ecosystems, or for the value & importance of non-human life.

Sorry to say, but Main Street's buying choices often are not helping the situation – petrochemical fertilizers & pesticides, single-use plastic bottles, many other forms of plastic, fossil fuels for nonessential uses (e.g., four-wheeler offroad recreation, gas-powered lawn mowers, monster-truck recreation, etc.), clothing made from petrochemical fibers (nylon, dacron, orlon, and now, *plastic*), and of course, large gas-guzzling automobiles. Never mind all the goods which essentially are temporary consumer "junk", and contribute to our ever-increasing production of wastes which often are dumped into the soil, the air, or the water.

That all must change, and soon, and massively... **a paradigm shift**. All of us can help by seriously re-evaluating our ethical beliefs regarding existence and beings. A shift to the holistic worldview of ecoethics is absolutely essential. Such a change must come from the ground up in society, not from the top down.

Education is key to the Big Picture. At every level of education, we must work to infuse the system with awareness of an ecoethical worldview. Faculty and students should be aware of it as a possible choice, a viable option regarding a philosophy of life. That means going beyond just teaching ecology, or touting the benefits of "going green". It means digging into the discipline of philosophy, particularly, ethics. It means a type of classical and liberal education, not merely a technical education. It's practically the opposite of what's going on today in most schools. It's a type of education which produced the great minds in our somewhat distant past.

Decades ago, Bucky Fuller wrote, *"Specialization precludes comprehensive thinking."*. I think he was right. In general, what has been promoted in education since then?   Of course, specialization. As a consequence, the world seems to have arrived at a point where very few people can think comprehensively. One result has been the present socio-eco-econ-ethical Crisis in which we find ourselves.

At the same time as offering ecoethics in schools, or if it's a pipe dream to believe it could be brought about in formal education, do it informally. At least in developed countries, most people have access to the internet... the world's greatest library. Every one of us could educate ourselves and/or our children in an ecoethical worldview...

basically at no cost. Are you serious about wanting to help save organized human existence, or not? It will take work.

NOTE –

None of the preceding in this section are solutions to our current Crisis. Rather, they are precursors or forerunners to the successful implementation of various solutions. If we're still lost in consumerism/materialism when we attempt to implement solutions, most likely we'll fail at that task. [Specific solutions are found in Chapter 11.]

Chapter 3: Climate Disruption

Where We Stand Now

Nothing demonstrates mega corporate control of politics like our inadequate response to the ongoing climate crisis. *At the national level,* we're still tinkering around the edges of the problem. One could say generally: lots of talk, but little truly significant action in most regions around the world. Most progress toward mitigation of climate disruption has taken place at the State and local levels of government. Of course, that's my considered opinion as an ecologist. Other people may disagree.

From 2019 through half of 2021, about 14,000 scientists from over 150 countries added their signatures to a document declaring a worldwide climate emergency. Hopefully by now, most of the world has resolved the question, "Is it really a crisis?". For decades, the evidence has been mounting that *even if* we aren't the sole cause of Climate Disruption, we certainly are a significant cause of it, and overwhelmingly so. In either case, we must drastically reduce the amount of greenhouse gases (GHGs) we're dumping into the atmosphere if we are to avoid an unprecedented catastrophe. It's just common sense. And, tick-tock.

The Intergovernmental Panel on Climate Change (IPCC) recently released its latest Report (August 2021). Hundreds of scientists from around the globe analyzed over **14,000** scholarly research articles on climate. They concluded that we are in dire need of immediate mitigating action, and that there's no longer any question about the cause of climate disruption – it's human activities. Their last Report prior to this new one was about ten years ago. According to their study, if we don't act now, an utter catastrophe will hit us no later than 2050. The way things are going presently (re heat domes, droughts, floods, wildfires, loss of glaciers, much more intense storms, deaths, etc.) I believe that year should be changed to about 2035.

Some recent studies show strong evidence that due to "feedback-loops" involving melting ice and the amount of the Sun's radiant energy reflected versus the amount absorbed, even if we drastically reduce GHGs the planet will continue to warm for years and years. According to those studies, we must not only reduce emissions but we must also capture and "sequester" carbon on a massive scale. That may never happen on a scale large enough to be truly significant. In other words, it's too late to completely avoid all climate disruption; however, if we do essentially nothing, the consequences of our inaction will make things much worse than if we try our best to solve the problem.

Some people online have wondered what they'll do when the climate crisis hits, how they'll adapt. All they need do is take stock of what they're doing *now...* because <u>that crisis already has started</u>. Consider these points:

1) Earth had an estimated net loss of 28 trillion tons of ice between 1994 and 2017;

2) each year on average, the southern edge of the Arctic's *permafrost* region moves North by about 1.0 to 1.5 miles;

3) as permafrost thaws, it releases methane gas to the atmosphere (and methane is multiple times more powerful a GHG than carbon dioxide); and

4) the number of internal-combustion engines has increased exponentially in the last 100 years, and when operating they produce nitrogen oxides, some of which are more powerful a GHG than is methane.

5) the Sahara Desert has increased in size by 10% since 1920.

The Mindset of Politicians and the Bane of Bureaucracy

In October 2021, the Transnational Institute report, *Global Climate Wall - How the world's wealthiest nations prioritise borders over climate action*, revealed that the governments of the USA, the UK, Canada, Australia, and several others spent over twice as much on border security as on mitigating the climate crisis. Shortly after the report was issued, COP26 was in session in Glasgow. The climate activist, Greta Thunberg, referred to those proceedings as a "failure… a Greenwash Festival". Other critics denounced the final COP26 Pact because it weakened earlier Drafts which called for the phasing out of fossil fuels.

In general, politicians love the status quo. They rarely ever have any desire to promote and work for a major change. Why? I suspect it's because each of them has one main priority while in office: to get re-elected. That means each of them is loathe to rock the boat.

So, what does it take to get them to do so? History repeatedly has shown that it requires massive, peaceful pressure from Main Street. In most cases, politicians will not risk upsetting the "average" voter unless said pressure is applied. Keep in mind, our "representatives" meet with paid, *special interest* lobbyists just about every week of the year. Those lobbyists not only hard-sell their point of view, but also often present the politician with a first draft of some legislative bill or proposed amendment to an existing law. That's our competition, and it's dedicated. Hundreds and hundreds of lobbyists from the fossil fuel industry, the petrochemical industry in general, Big Pharma, the factory farming industry (Industrial Ag), the transportation industry, and others who most likely are opposed to a holistic view are working hard to maintain the status quo. Such work will not serve us or Spaceship Earth well. It must be countered.

In terms of the climate crisis, this all means for any significant mitigating action to come out of DC, our lobbying efforts (so to speak) must be undertaken by large numbers of people. Persistently. If marching in the street is not your thing, there are many other ways to have an impact. Here are only a few.

1) Educate yourself. Do an online Search to discover which legitimate groups are promoting policies for the mitigation of climate disruption. Pick one, and support it. Every little bit of financial help counts. It's the very least each of us can do.

2. As best you can, don't patronize companies-brands-products which seem to act as if the atmosphere is a free dumpsite for their air pollutants. That would include companies/industries which lobby against air pollution regulations. Some of them claim that such regulations are "socialism", but having clean air is merely common sense. This item requires a bit of research on your part.

3. If you are a writer, a painter, a photographer, a teacher, an entertainer – whether amateur or professional – when appropriate, infuse your work with climate education.

4. Do some research on the agendas first, and then attend local school board meetings to suggest means of climate education. Or at least discover if any is present in school

curricula by meeting with or calling school staff. Since time is relatively short regarding the Crisis, this item is especially crucial at the high school and college levels. This is not flashy or dramatic work, but it can have an impact. We need young people who have more education in this realm than just media sound-bites.

5. **Utilize social media** to promote the solutions found in Chapter 11 of this book.

NOTE - Again, the above suggestions are not solutions, but rather, forerunners to the implementation of solutions to our socio-eco-econ-ethical Crisis.

The Ocean and Climate Change

Our ocean plays a central role in the regulation of both weather and climate. The temperature of major ocean currents greatly influences them both. With the ocean absorbing more and more carbon dioxide from human activities, ocean waters are warming and acidifying. Temperatures are increasing as far as 3,000 feet down. One scholarly study predicts before 2080 there will be significant deoxygenation of large areas of the ocean (see Gong, Hongjing et al. In "References"). Why? Warmer waters hold less dissolved oxygen gas. When CO2 dissolves in water, carbonic acid is formed. More and more of it results in ocean acidification. Both warming and acidification of the ocean are two reasons climate change is inextricably linked to biodiversity loss.

Not only does the ocean affect weather and climate, but the atmosphere affects the ocean in another way as well. When CO2 increases in the air and is absorbed into the ocean, the resulting carbonic acid slowly starts dissolving the skeletons of coral and the shells of some sea animals. Key organisms are negatively affected. Food webs are damaged, entire ecosystems are damaged, and biodiversity is reduced.

Chapter 4: Humans and Materialism

Each of us conducts a search for contentment and real happiness. For most people, I suspect it's a genuine struggle which occupies much of a lifetime. During that journey, we form a philosophy of life. As part of that philosophy, we often purchase buffers against despair: a new car, a large TV, more clothes than a person needs, and all manner of gadgets and devices. Some turn to alcohol or other kinds of drugs. Such contrivances, though, offer only temporary insulation; hence, we are never satisfied.

The Philosophical Problem

We've all been sold a false bill of goods regarding a philosophy of life. From an early age, we were indoctrinated into the belief that more is better. More and more material goods supposedly will make us all "happy", and society overall will benefit to the maximum. It seems almost everyone in our lives has believed that falsehood, and passed it on generation after generation. The story goes like this: if each of us can get to the stage of being materially (and financially) "well-off", then we'll all have meaningful, fulfilled lives. It's often presented here in this country as "The American Dream": your own home, at least two cars, plenty of clothes, appliances, money, etc.

Unfortunately, it simply isn't true. Meaning, fulfillment, and happiness all are in a sphere which has almost nothing to do with materialism. Obviously, it's a given that we do have certain basic, material needs – air, water, food, clothing, and shelter. But beyond those items, don't expect more and more to make you happy. More of everything sometimes can bring you convenience, a higher degree of physical comfort, and even luxury in some cases; but none of that is happiness. Mostly, it's merely comfort, and primarily, physical comfort. There's nothing wrong with that, but it has little to do with having a meaningful and fulfilling life.

Also unfortunately, endless pursuit of more and more has dire, unintended consequences, as follows.

1. Any satisfaction we attain is short-lived. Before long, we're back to grasping for even more because the supposed "happiness" from the last batch of things has evaporated.

2. The result is that we become frustrated and unfulfilled in our lives.

3. On top of that, in order to get more and more, most of us then fall into the often described "Rat Race". It's never-ending. It's debilitating.

4. Finally, more and more consumption requires an economic system which generates more and more "throughput". The outcome is more and more extraction of natural resources, more and more manufacturing, more and more environmental/ecological damage, and ultimately, *unsustainability*.

All the above is common sense, right? If so, why do we continue on such a destructive, unfulfilling path? A big part of the answer follows. The so-called Establishment or Powers-That-Be keep filling our minds with two key ideas.

1. Being well-off really does bring happiness… really, really, really.

2. Human genius and technology will solve any social and/or ecological problems generated by perpetual growth and overconsumption.

[We make a serious mistake when we believe either one of those assertions. Look to your own experience and to history for confirmation of the mistake.]

Fallacies of "The American Dream"

The first mistaken belief found in the "Dream" concept is the implication that more and better goods/personal infrastructure will result in happiness and fulfillment. The second fallacy is that regardless of one's socio-economic status, working hard will win you the prize. In other words, because the playing field is level, all you have to do is work hard and perhaps have just a tiny bit of good luck. Put another way: hard work guarantees upward mobility. The third fallacy is the implication that this all happens only (or maybe primarily) in the USA. Why? Because we supposedly live in a free democracy, and are exceptional, and the envy of the world. The reality is much different.

Yes, we live in a great Land. We are unique as well, but many other countries are every bit as great and unique as ours. Here's the problem with the idea of "American Exceptionalism": there's a whiff of arrogance in that concept. There's an air of superiority and "holier-than-thou". Unfortunately, with some people it becomes toxic.

Perhaps more importantly, hard work does not necessarily mean someone will move up the economic ladder. Why? It's because *the playing field often is not level for everyone.*

The Powers-That-Be sometimes play favorites. There are so many examples of such in the News and in the experience of most people every so often that listing them is not necessary.

True Wealth

The possession of inordinate wealth resulting in over-the-top physical comfort is an important goal for some people, but ultimately, it becomes unsatisfying. *It's never enough.* More and more is needed, and that's not enough either. Such is the reality of a life mired in materialism which is way too far beyond basic necessities. It's okay to go a bit beyond the basics, but too far and it will never end for too many people. I've lived a relatively long life and have seen it in people more than a few times. We've all seen examples of it in various documentaries or News reports about unhappy, disturbed, or frustrated rich people. When it comes to material goods, some version of minimalism is best. It frees one up to engage in meaningful, satisfying endeavors.

True wealth is found in family, friends, health, purpose, peace of mind, and an omnipresent connection to Nature. Add to that: contentment, equanimity, and as Eckhart Tolle puts it, *the radiant joy of Being.* The experience of those is far more valuable than the latest trending gizmo. The more material goods you own, the more they own you. Ultimately, it's a path to dissatisfaction and frustration. Ignore the Edward Bernays style propaganda which says otherwise.

Chapter 5: The Nuclear Question

A Bit of Background

Henri Becquerel accidentally discovered spontaneous radioactivity in 1896. He didn't realize at the time that it came from undiscovered chemical elements. Along with her husband, Pierre, Marie Curie did numerous experiments in the early 1900's on radioactivity. They worked with the mineral, pitchblende, and discovered two new elements (naming them radium and polonium). In 1903, the Curies shared a Nobel Prize in Physics with Becquerel. Marie Curie also won the Nobel Prize in Chemistry in 1911 for isolating pure radium in the lab. She died from leukemia in 1934.

Not long after the discovery of radium, companies started adding it to all sorts of consumer goods. There were dinner plates, lipsticks, tonics, cocktails, paints, toothpaste, animal feed, watches, and other products which all contained the radioactive element. People loved the glow, and many believed radium had curative properties. Various radium tonics were billed as "liquid sunshine", and guaranteed to be "completely harmless". Even a scientific journal, *Radium*, claimed in 1916 that the element was "harmless".

When I was a kid in the late 1940's and early 1950's, there still were radium dials on watches, or the hands of the watch were coated with radium paint. In *shoe stores* there were vertical boxes (similar to a podium) with a viewfinder on top. A person would stand on an attached platform, look through the viewfinder down to a large slot at the bottom in which your feet were placed, and you could see the bones of your feet and toes. Basically, it was an X-ray machine. A few of those still were in use as late as 1970!

Nuclear Weapons

I was about two years old when the atom bomb was developed during World War II. The Head of that project, the American physicist J. Robert Oppenheimer, had reservations about the use of nuclear weapons. The first test detonation of the A-Bomb (a fission bomb) was in July, 1945 in New Mexico. In 1949, the old USSR tested its first A-Bomb, and thus the nuclear arms race was in full swing.

Thanks to the work of Edward Teller and others, the first hydrogen bomb (a fusion bomb) was tested/detonated by the USA in May, 1951. The H-Bomb is at least four times more powerful than the A-Bomb. In March, 1954, a more powerful H-Bomb was detonated by the USA in the central Pacific ocean (Bikini Atoll, Marshall Islands). If you've seen the video of that explosion, I'm fairly certain you would agree that the H-Bomb is an obscenity. From 1945 until September, 1963, **hundreds** of nuclear bombs were test detonated **above ground** by the USA, France, the USSR, China, and the UK. Succeeding bombs usually were more powerful than previous ones. Early on, our Government assured us there was "no appreciable danger" in above ground testing. It wasn't until 1959 that the Feds finally admitted nuclear fallout was dangerous to humans. Nevertheless, above ground testing continued. Meanwhile, in the early to mid-1950's, nuclear electrical power plants were being pioneered in both the USA and the USSR. Back then, it was thought that nuclear power soon would be providing free electricity to everyone.

Nuclear Wastes

No one gave too much thought to nuke waste at the beginning of the nuclear era. In the 1950's, small amounts were put into 55-gallon drums and buried or placed in pools of water… as an interim storage solution. Larger amounts were placed in huge storage tanks, first single-shelled and later double-shelled. The 55-gallon drums soon were replaced with steel and concrete "casks". The Fed Government tried to mandate a few permanent geologic storage sites in Kansas and Nevada over the years, but all were eventually terminated because of tech uncertainties or local opposition, or both. The Yucca Mountain repository site in Nevada was proposed in the 1980's, and all work on it ended in 2011. It never came to be. Incredible!

So, you might ask, where are the tons of radioactive waste being stored currently? The answer is: on whichever site they are generated. Three-fourths of all States in the USA have such sites. All of the storage is *temporary* until the Feds locate a permanent site and get it built and operating. We have been generating nuclear waste since 1945, and still don't have a permanent repository for it. The high-level waste remains radioactive for thousands of years.

For almost two decades just prior to 1975, Dow Chemical - a private contractor - operated the Atomic Energy Commission's Rocky Flats bomb factory in Colorado. Because of Dow's improper storage of thousands of drums of oil contaminated with

plutonium, those drums corroded and leaked plutonium into the soil, water, *and air*.
Dow also "disposed" of contaminated water by spraying it on fields surrounding the
facility...claiming that was "irrigation". A Whistleblower by the name of James Stone
brought this all to light, and of course, he was fired or forced to resign. In 1975, a new
Government Department took over and a different contractor was hired.

Throughout the 1980's and culminating (in 1989) in an FBI raid on the Department of
Energy (DOE) Rocky Flats nuclear bomb manufacturing facility, Rockwell International
(the other private contractor) was illegally disposing of hazardous waste contaminated
with plutonium. Plutonium has a half-life of 24,000 years and is considered by many to
be the most toxic substance on Earth.

The illegal disposal resulted in plutonium contaminating not only soil, but the air, the
groundwater, and surface streams as well. Over sixty pounds of the toxic substance
were found to have accumulated in the ductworks of just one building of the plant. The
US EPA and the Colorado Dept. of Health both were prohibited by DOE from inspecting
the key areas of the entire facility. "National Security". Meanwhile, plutonium was
contaminating the air, soil, and streams that wound up in drinking water reservoirs.
Eventually, DOE succumbed to pressure and submitted a huge Report detailing the
hazardous waste and what was done with it. The FBI studied all 27 volumes of the
Report, and even took water samples just offsite of Rocky Flats. That all led to the raid
on the facility...one Agency raiding another.

There's more to the story, going into the 1990's. If interested, go online to YouTube and do a Search for "Rocky Flats"; you'll find several documentaries or clips. In particular, look for *PBS Frontline*'s--- *Rocky Flats Secrets of a Bomb Factory.*

In the 1970's, I took an Advanced Ecology class of mine on a trip to visit the Hanford Site in the State of Washington. Back then, it was a nuclear production facility covering 586 square miles. *In 1943, Hanford was the world's first plutonium production plant, and part of the A-Bomb Manhattan Project.* It generated a lot of high-level nuke waste over the years. At the end of the Cold War, it was decommissioned. Today, it's the most contaminated nuclear site in the United States. When it closed down, it left behind **53 million gallons** of liquid high-level radioactive waste and 25 million cubic feet of solid nuke waste.

In 1989, Hanford became the largest environmental/Superfund cleanup site in the country. In 2011 and 2012, it was discovered that several of the large tanks were leaking nuclear waste into both groundwater and surface water. During its operating years, the site used Columbia River water to cool its reactors. All of this resulted in nuclear contamination being detected over 200 miles from Hanford.

When my class and I were there, of course, we were not aware of any of these problems. We were given the typical tourist-type tour, and were impressed with how spiffy and clean everything appeared to be. So much for appearances.

ICAN

From their website: "The International Campaign to Abolish Nuclear Weapons (ICAN) is a coalition of non-governmental organizations promoting adherence to and implementation of the United Nations nuclear weapon ban treaty." In 2017, ICAN was awarded the Nobel Peace Prize. In 2021, nuclear weapons became illegal under international law. So far, 59 countries have ratified the treaty. The USA is not one of them. Neither are the other eight countries which have nuke weapons. That's no surprise, but arguably inexcusable.

The good news is that the days of such weapons being viewed by Main Street as acceptable and necessary are fading into history. The overwhelming majority of people around the world want them banned. Not only are nuclear weapons a threat to all life on Earth, but according to several high-ranking officers in the military (or retired from active duty), they are not necessary for the defense of any nation. One of those retired officers is Colonel Andrew Bacevich (see "References"), who is also a Professor Emeritus of International Relations and History at Boston University.

Chapter 6: Renewable Energy

The Big Picture Problem

People on Main Street, in government, in academia, and in business/industry all are encouraged by the thought of renewable energy, and rightfully so. Except for many in the fossil fuel industry, most people appear to realize we must transition away from fuels derived from petroleum. Problems associated with those fuels are fairly obvious to almost everyone.

Consequently, "Going Green" seems to be the order of the day in many circles. Many institutions (public and private) are taking that shift seriously. Some (mostly private) appear to be taking the window-dressing approach. In any case, progress is being made in various locations around the globe. Some of it is significant. **But, wait; there's a flaw**.

Most of the Green proposals (at the macro level) are various versions of what's called the "circular economy". In that approach, emphasis is on renewable energy, durable goods, reuse, regeneration, recycling, using waste as a resource, and using fewer natural resources. Energy and resources are in a circular rather than linear flow. Sounds great, but all the proposals seem to retain the theme of *unlimited economic growth*.

Many, or perhaps most, ecologists agree that Going Green while maintaining unlimited growth is not feasible. The previously discussed elements of entropy and biophysical constraints come into play. Whether Green or not, in a finite habitat eventually growth has to be limited or the system collapses. So, what's to be done? We'll get to that in Chapter 8, "Ecological Economics", and Chapter 11, "Solutions and the Future".

Another problem regarding the big picture of energy: it appears too little attention is being given to *small scale*, self-contained, household and small business level means of implementing renewable energy. To an overwhelming degree, the emphasis seems to be on the Big Energy industry and its massive grids. A few struggling companies are working diligently on developing small gizmos which can be used by individual buildings to obtain energy. How much government support are they getting? Any subsidies? How about tax breaks? Any help at all?

I imagine that Big Energy is opposed to any such development. Except for large facilities, small scale energy machines would drastically cut into their business. In many places, the Big Boys/Girls already are giving homeowners a hard time relative to power from rooftop solar panels. And Big Energy has a powerful lobby in governments.

Finally, a few astute scientists have put forth the idea that we don't have much of an energy problem; instead, we have a **consumption problem**. Especially in the developed nations, we use too much energy. That seems to be driven by unlimited economic growth and overconsumption of goods. Both of those are <u>unnecessary</u>. A paradigm shift in ethics and worldview would help tremendously in changing the situation. What must be overcome is the Edward Bernays style propaganda which convinces too many of us that to be "happy" we need to be gluttons relative to all kinds of goods.

The Biofuels Problem

There are a few problems with biofuels, but the biggest one is this: relative to the generation of electricity, the current emphasis is on the burning of wood. Trees are converted into wood chips and burned in power plants. Here are the reasons why that's not a good idea.

Trees capture carbon dioxide and convert it into glucose and oxygen. Burning wood chips (trees) adds carbon dioxide to our atmosphere while reducing the carbon fixation capability of the forest by harvesting trees for such burning. We have too much CO2 already in the air. Trees increase an ecosystem's biodiversity, help prevent soil erosion, and add humus to the soil, thus increasing its fertility. According to the botanist, Diana Beresford-Kroeger, many trees produce aerosols which have antibiotic and anti-cancer properties. When you're in a forest, you inhale those and boost your immune system. In my opinion, trees should be revered, not ground/chopped into wood chips and burned as some sort of "Green" solution to a power plant fuel problem.

Another biofuel problem is ethanol from corn. It takes more energy to produce a quart of ethanol than the amount of energy found in that quart of fuel. The whole process of production and distribution is energy-negative. Corn is better suited to being used as food for humans and some nonhumans. Currently, about 40% of USA's corn production is for the manufacture of ethanol. The price of corn has gone up because of demand, resulting in somewhere around a million acres of wildlife habitat per year being converted to corn fields.

Even with other sources of potential biofuel there are significant problems not yet resolved. Sugar cane and cellulose from switchgrass are prime examples. The main problem is the cost of producing the fuel. It's way too high. Even with its high cost, biofuel from algae (pronounced "algee", with a soft g) is the overall best choice. You'll see why in Chapter 11, "Solutions, and the Future", subsection "Energy".

The Wind Turbine Problem

In a nutshell, these marvels of engineering are too large, use too many resources in both manufacturing and installation, and are disruptive to wildlife. In addition, the noise they generate can be disruptive to humans in serious ways. You don't want to live anywhere near a wind farm. Although not yet fully developed, smaller wind devices with few-to-zero moving parts are being researched. Whenever possible, governments should do what they can to support such efforts. In my opinion, they'll be needed.

The Solar Energy Problem

The main issues with solar panels are: low efficiency; the resources their manufacturing requires; and their life-span. Improvements are in the works, but solar panels *alone* never will be sufficient enough to handle the world's power consumption. Even with ongoing improvements in batteries for solar energy storage, additional other renewables will be required to meet the demand. I mention this only because it seems some people believe solar energy is the silver bullet which will solve the energy problem. It's not. Other sources will be needed as well. Despite the above problems, progress on renewables is continuing. Their future appears to be bright.

Chapter 7: The Land Use Dilemma

Expansion of the Human Footprint

As human population increases, more and more farmlands and wildlands are lost to residential, commercial, industrial, and governmental (e.g., roads, highways, airports, dams, etc.) development. Worldwide, massive changes to forests, rivers, wetlands, prairies, and other habitats have increased biodiversity loss, soil degradation, the shrinking of forests, and the loss of (or damage to) other natural ecosystems. These changes are not natural or sustainable.

Because natural ecosystems are dynamic, they undergo almost constant change. The difference between that change and the ones mentioned in the paragraph above is important. With a few exceptions (such as wildfires caused by lightning, volcanic eruptions, a large meteor striking Earth, or huge landslides), changes caused by Mother Nature generally occur *slowly over time*. For example, through an ecological process known as "succession" a sandy beach may be transformed into an area covered by grasses and shrubs. Or a wetland may be converted into a forest. Plants, animals, and microbes have time to gradually adapt to the changing circumstances.

Changes wrought by us usually are a different story. In many/most cases, they occur relatively quickly and with damaging effects on natural ecosystems. It started millennia ago with the advent of agriculture. Although, the effect back then was mild compared to today's industrial agriculture, strip mining, suburbanization and associated sprawl, clearcutting of tropical and boreal forests, and industrialization in general. Despite our efforts in conservation, for the most part too many of our activities have wreaked havoc upon our life support system, Nature.

Loss of Natural Habitat

With the expansion of our ecological footprint came the increasing demand for water, wood, minerals, and energy. To meet the demand, our impact on natural habitats all around the world has been devastating. That must change if we are to survive and thrive.

The Dilemma

Human existence brings about essential human needs, and as our numbers have increased, so too has our extraction or appropriation of natural resources in order to meet those needs. Put simply, we have to use land (as well as other resources) for various purposes – agriculture, manufacturing, housing, forestry, mining, and the like. Unfortunately, history has shown that too often we overuse or misuse land, air, and water...and the organisms on or in them. That's the dilemma: how do we continue to use land (in this case) for our essential needs *without degrading it to the point where our natural life support system is ruined*?

There are numerous answers to that question. Many ecologists (including myself) believe the main key to improving the situation lies in adopting and adhering to some

form of ecoethics. Such will result in the dedicated and serious pursuit of global sustainability. The current status quo will not suffice.

Chapter 8: Ecological Economics

Preface and Principles

If we are to avoid an ecological catastrophe, then a relatively rapid paradigm shift must occur in a number of key areas. One of those areas is economics, specifically, the neoliberal/neoclassical economic system which currently is dominant worldwide. In my opinion, here's why the shift should be to ecological economics (EE).

For two-plus decades, I've been puzzled (baffled, really) as to why neoliberal thinking views the economic system as *separate from Nature*. Any economic system is embedded in the natural world - in ecosystems, in biogeochemical cycles - and is subject to the Laws of Nature, so to speak. Do we not extract timber, other plant products, common minerals, rare earth minerals, water, fish, etc. from the natural world? Do we not all breathe air? Are not all businesses (and consumers) sometimes subject to the whims of flooding, storms, droughts, natural vectors of disease, and the like? Obviously, the answer to each question above is YES. Consequently, it makes no sense to me for anyone to believe that any economic system is not a smaller piece of the natural world. In a very real sense, even though this planet has innumerable ecosystems, Earth itself is one giant ecosystem. No artificial system is separate from or larger than that.

One of the founders of ecological economics, Herman E. Daly, has discussed the concept of "illth" (as opposed to wealth) in depth. Illth refers to the unaccounted downside ("negative externalities") of unlimited economic growth. Examples would include: nuclear waste, toxic pollution, excessive deforestation, soil erosion, climate disruption due to greenhouse gases, biodiversity loss, urban congestion, the "rat race", etc. Mainstream economists seem to believe that GDP is an accurate measure of well-being in a society, but they ignore the illth which results from an ever-growing economy. A higher GDP results in a greater generation of the ignored downside. At some point, illth decreases our well-being, thus higher and higher GDP produces *uneconomic growth.*

William Rees, founding member and past president of the Canadian Society for EE, wrote a brilliant piece several years ago for the Great Transition Initiative: https://greattransition.org/publication/economics-vs-the-economy In that essay, he makes the point that EE recognizes the economy as "an open, wholly dependent subsystem of the ecosphere...". Natural resources are extracted from, and wastes are injected back into, Nature. Any material transformations in these processes are subject to Natural Law (e.g., the Laws of Thermodynamics). In other words, any economy is not outside of or separate from the natural world.

The neoliberal economic paradigm operates as if the opposite were true. As a result, our world is now mired in a plethora of ecological disasters: loss of biodiversity, which is damaging ecosystem services to humanity and nonhuman life; habitat destruction causing (among other things) disease proliferation; crucial ecosystem damage (e.g., to wetlands & their critical functions of flood control and removal of toxic substances from water); acidification of the ocean; the sixth mass extinction event on Earth; air pollution; water pollution and unsustainable use of water; climate disruption; and more.

EE also takes into consideration the concept of ecological constraints on what Kenneth

Boulding in 1966 and Bucky Fuller in 1969 called "Spaceship Earth". Except for incoming solar energy (and some meteorites), we live in a finite, closed habitat. It's not growing larger. Fortunately for life here, there are many, many ongoing regenerative and

biogeochemical cycles of materials on this ship. When we interfere with those cycles beyond sustainability, we threaten our life support system's proper functioning. When we do not properly manage industrial wastes, and when we overharvest natural resources beyond sustainability, essentially we are committing species suicide. EE recognizes these problems, but neoliberal economics mostly ignores them.

In addition to those mentioned above, key elements of EE include the following:

1. a transition to a *dynamic* steady state (not stagnant), sustainable economy rather than the current unlimited growth model;

2. the recognition, incorporation, and evaluation of natural capital and ecosystem services in the economic system;

3. efficient allocation of resources, fair distribution of goods & services, and a sustainable scale of the macroeconomy (scale = the physical volume of the throughput);

4. sustainable development;

5. local/regional procurement of goods & services whenever possible;

6. reduction of material "throughput" in the economy;

7. a transition away from fossil fuels; and,

8. the incorporation of ecoethics into economics and the economy in general. [In my opinion, EE does not focus on this element nearly enough; however, it does so much more than mainstream/neoliberal economics...orders of magnitude more.]

As we navigate through the Anthropocene geologic time period, we must face up to a few facts that we've been avoiding.

1. As Rees and others have pointed out: the finite ecosphere in which we live has highly variable, but limited, regenerative *and waste assimilating* capacities.

2. Our current economies around the globe and all the activities associated with them are destroying significant sections of our natural life support system.

3. That destruction primarily is due to the pursuance of unlimited economic growth, un-limited land development, and overconsumption of goods.

4. The main driver of the above pursuits essentially is unlimited population growth.

5. Despite some positive applications, our technology has not slowed the pace toward impending, massive ecocatastrophe. Arguably and overall, it instead has increased that pace. [Many believe that's especially true of the High Tech Industry. Some of those companies now claim to be 100% "Green". They can do such because they contract out the manufacture of goods/parts to *other companies*, and those firms generally are the ones with poor environmental records.]

6. To avert more disaster, we are in need of a rapid cultural and ethical evolutionary change in our thinking and behavior.

7. A significant part of that change should be the shift away from neoliberal economics.

8. After much examination of the factors involved, I believe we should adopt and imple-ment some version of ecological economics in as many countries as possible. That would be a big step toward improving and preserving organized human existence on Spaceship Earth.

Conflict

EE is a transdisciplinary science. Not only are ecologists and economists involved, but general biologists, sociologists, physicists, geographers, and others are as well. Different viewpoints are encouraged. Over the years since the founding of EE in the 1980's, somewhat of a split has developed among the scientists in this academic field.

Those who adhere to the original ecological approach to EE have expressed concerns about others who appear to be leaning toward mainstream (neoclassical/neoliberal) economics. Just for the sake of illustration or discussion, let's call the first group the "originals" and the second group the "co-opters". To varying degrees, the originals feel that the Ecological Economics Movement is being co-opted by those who take on the color of Green, but only within the framework of mainstream economics. In other words, they feel that EE is gradually being diluted by the co-opters to the point where soon it will be unrecognizable. In their view, EE must return to its roots.

Such roots go back to the pioneering works of Nicholas Georgescu-Roegen (viewed as the Father of EE) and his 1971 brilliant text, *The Entropy Law and the Economic Process.* [See "References"] One of his PhD students in Economics at Vanderbilt University in the mid-1960's was Herman E. Daly. Daly [See "References"] went on to co-found the International Society for Ecological Economics in the late 1980's, and was a co-founder of their scholarly Journal, *Ecological Economics*. The Journal still is *the* premier source for information on EE research and implementation.

Another "split" (of sorts) in the EE world revolves around the difference between Georgescu-Roegen's vision and that of Daly. The former emphasized the role of entropy in macroeconomics, and maintained that even a Steady State economy cannot exist forever in a finite habitat. Daly agreed, but still thought his Steady State model was the best choice for society. Georgescu-Roegen advocated a Degrowth model. The disagreement largely is something akin to splitting philosophical hairs, and could be resolved simply by joining forces and promoting degrowth *to* a steady state economy. Such has been suggested by more than one analyst.

Those disagreements help explain why "sustainability" is taking so long to be adopted. [Chapter 11 goes into it even more.] Main Street may or may not be aware that before a theoretical concept becomes implemented in practice it needs strong philosophical underpinnings. Without those, implementation has a good chance of failing.

The Steady State Economy

Books have been written on this subject. For readers wanting specific, detailed information, simply do an online Search for same. Due to time and space constraints, all we can do here is provide a few brief and general observations about the theory, as follows. Characteristics of the Steady State Theory---

1. A Steady State Economy (SSE) is NOT a "Nanny State" (not even for corporations) or an authoritarian socialist state.

2. Given that humans do require natural resources to survive, as much as possible SSE allows Nature to heal from the current damage inflicted upon Her.

3. By voluntary, noncoercive, and humane policies and incentives (See Chapter 9), SSE advocates lowering human population on Earth over time.

4. SSE favors innovation and sustainable development rather than unlimited economic growth and overconsumption.

5. It also favors *public* banks. National governments should create new money, not private banks. Basically, it would be debt free. [For those who may not know, private banks create money via "fractional reserve" banking and making loans.] Also, state owned banks are much more accountable than private ones. [Do an online Search for "Ellen Brown on banking".]

6. SSE favors a Universal Basic Income for all citizens. Or, call it a Universal Basic Dividend. Such would be much more efficient *and cost-saving* than the current welfare system. Plus: it would essentially eliminate almost all fraud in that sphere; would help ameliorate our gross income inequality and social inequity; and it would likely be pumped right back into the economy.

7. SSE favors policies promoting local and regional economies. Such economies would strengthen supply chains, make manufacturing facilities (and other businesses) more accountable to customers, and give more consideration to people in economic planning.

8. It also favors Small Business, the backbone of any successful economy.

9. SSE fosters a balanced Mixed Economy of market capitalism and democratic socialism.

10. It also promotes ecosystem protection and an evaluation of the services ecosystems provide to humanity and all other life.

None of the above means SSE is against extracting natural resources from ecosystems. We have to do that in order to support and enhance our lives. It's a matter of doing it sustainably. Our current economic system fails in that regard.

Awareness and You

Unless you're in the top 15-20% of income earners in your country, you may feel trapped in the proverbial Rat Race. You may feel that's just the way it is, and there's not much you can do about it. Perhaps you've bought into the propaganda which claims the economy must continually expand, and must be mostly global, and must be run by elites, no matter what the cost in human contentment and environmental damage. If so, you probably would benefit by doing a bit of research on alternatives. In short, become aware. Other choices exist, and they are feasible or at least possible.

Economics is a human construct. No one type of it is carved in stone; nor is it any kind of Natural Law. Don't believe for a moment that our current economic system has proven itself to be the best for the world. That's propaganda of the worst sort. Neoliberal (aka, Neoclassical) Economics is the best only for the Upper Crust. Anyone who really looks at the condition of the world today cannot but agree with such an opinion. To widely varying degrees, everyone below the top 15-20% or so is suffering. To make matters worse, nonhuman life and entire ecosystems are as well. A big part of the cause of that is neoliberal economics, our current economic system.

Be aware: SSE can change that situation for the better...much better.

Chapter 9: The Elephant in the Room

Human Population

Despite human numbers decreasing somewhat in a few areas around the globe, population growth on Earth continues mostly unabated. We're now at about eight billion, and rising. [NOTE: Larger increases in poorer countries (as compared to richer nations) have <u>nothing whatsoever to do with Race</u>. It's a matter of a family's socio-economic status; poor people and those without much education generally have more children than those well-off and educated.] According to many ecologists, we exceeded Earth's carrying capacity years ago. Simply put, human population growth likely is one of the most basic causes of our ecological and social Crisis. In a finite habitat, such growth has to level off sooner or later either by natural means or by such means *plus* non-coercive policies and practices instituted by us which promote lowering of the human population. Can anyone argue successfully against that thesis?

Worldview

For various reasons (religious, economic, ethical), some people argue against abortion, birth control methods, and any other policies/practices which would reduce human numbers. They appear to believe it's simply wrong to encourage such things. At the same time, I'm guessing they must have some sort of faith or hope that technology and the great ingenuity of our species *somehow* will avoid mass famine, pandemics, wars, water shortages, overcrowding, etc. caused by too many people. I've often wondered what the tipping point is for people with such beliefs. In other words, do they believe technology and imagination will avoid disaster up to the point of ten billion, or twenty billion, or maybe fifty billion of us on Earth at the same time? Is it unlimited in their view? Surely not, because that's not possible. To believe it is possible to keep growing in numbers with no serious adverse consequences is not only unrealistic, but also foolish. Surely we can do better.

In general, it would appear that people who are against reducing Earth's population fall into one of two "worldview" categories, perhaps with a bit of overlap in some cases. The first category is known (in some classification schemes) as the "Traditional" worldview. In it, reality is seen as theistic and dualistic. There's an all-powerful Creator God who is separate from humans and has a Plan for the Universe and all in it. Knowledge is gained through scripture, convention (such as schooling), and solid tradition. The values of such people include conformity, faith, discipline, and service. The second category is the "Modern" worldview, which sees reality as Newtonian/mechanistic, dualistic, and materialistic. Knowledge is derived from logic, the rational mind, and empirical science, while values include hard work, social status, success, power, and a hedonistic, materialistic existence.

The Traditional group respects Nature more than the Modern people do, because after all, in their minds Nature was created by God. To varying degrees, though, both groups engage in sometimes over-the-top exploitation of natural ecosystems. So what does all this have to do with human population growth?

It seems obvious that people with a Traditional worldview are against interfering with population growth because many or most of them believe such interference is a "sin", against God's will. They apparently get that from interpretations of scripture. When it comes to the Modern worldview, though, things get more speculative. I tend to believe it is a nearly obsessive interest in the economy and consumerism which drives the antipathy of those adherents toward any kind of human population management. Probably in their view, more and more consumers are needed to buy more and more goods. Such will enable the economy to basically expand forever, or so they believe.

Other people appear to be more realistic. They see the handwriting on the wall. In their view, it's simply common sense to humanely put the brakes on population growth. They tend to fall into either the "Postmodern" or "Holistic" worldview categories. Postmodern people view reality as both relativistic and pluralistic. Methods of attaining knowledge are (more often than not) qualitative rather than quantitative. Values include imagination, a love of Nature, and being open to change. The Holistic group views reality in a universal, non-dualistic, and evolutionary manner. Holistic people attain knowledge via synthesis rather than simple analysis, and believe in the integration of various methods of knowing. They value holism, wisdom, and spirituality.

So, relative to population growth, what's to be done, or rather, HOW do we do it.

Either Mother Nature or Humans Will Solve the Problem

If we don't adopt policies to reduce future human numbers, at some point in time they will be reduced on a massive scale by grim methods of Nature: famine, disease, lack of accessible water, drought, and the like. Sound familiar? It's only common sense to avoid more of those disasters by reducing human population. Following are a few possibilities for doing so.

1. Efforts must be doubled in terms of educating people, especially poor people, as to fertility, the methods of family planning, and birth control. It isn't stupidity but rather *ignorance* which abounds in many regions of the world.

2. After a couple has one or two children, a financial incentive/dividend could be offered the father to voluntarily get a vasectomy, or the mother to voluntarily get her tubes tied. The incentive could be a cash payment, or a tax rebate, or perhaps some amount of surplus food for a time, or even a small plot of land, or house repairs, or a free education for their child, or _____.

3. Women must continue to be empowered by education, not only education in family planning *but education in general*. Educated women have fewer children.

4. Some amount of universal basic income and essentially free higher education (at least a year or two) would help a great deal in reducing our population. People with a *steady* income and an education historically have had fewer children. There are always exceptions to a general rule such as that one, but they usually don't invalidate the rule.

5. Replacing the current neoliberal economic system would do much to enable workers to secure *steady* employment. Most jobs today are insecure to varying degrees. Many are part-time, or temporary, or both. Automation has eliminated a significant number of jobs. That will continue under neoliberalism. For example, it's predicted that by

2035 all transport trucks will be driverless, and truck driving currently makes up the largest single job category on Earth.

Years and years ago, neoliberalism claimed it would lift the entire world out of poverty. Its adherents may have believed that, but it has not materialized. Instead, the Rich have gotten richer, and most of the rest of the population is struggling. Many of those who were raised above the poverty line not only made it to only barely above that line, but also found themselves in a Rat Race with ever-decreasing job security.

It is believed by many ecological economists that a Steady State Economy would supply *steady* employment, which would mean a steady income. Smaller numbers of children per family would follow. Relative to a steady income, History agrees with that statement.

Chapter 10: Ecoethics is the Key to Surviving and Thriving

Is Human Extinction Possible?

Indeed, it is possible. 99.9% of all species that ever lived on Earth are now extinct. Of course, none of them had our intelligence or technology. Many people seem to believe those two things guarantee we'll go and on. Others are beginning to notice that while we usually are smarter than a pine tree, a dog, a monkey, and most all other life forms, we also are full of hubris, and appear to have lost the ability (or desire) to think comprehensively. Our arrogance has led us to believe we can exploit Nature and control it with few-to-no adverse consequences. A plethora of existential problems have arisen, and we're not dealing with them in an adequate manner. Any one of them likely could cause our extinction.

The Core of Ecoethics

Various scientists and writers (e.g., J. M. Ratcliffe, D. Suzuki, A. Leopold, D. Sparenberg, J. Lent, B. Commoner, John Cairns, Jr., O. Kinne, et al. - see "References") have described the need for ecoethics to be the basis of sustainability on Earth. Why? Simply put, it all boils down to two crucial concepts: everything is connected to everything else; and, planetary biophysical limits demand an ecoethical worldview. Without adherence to those concepts, sustainability essentially is most difficult to attain, and arguably, perhaps impossible.

In the metaphysical realm (the nature of Reality), western world culture accepts the idea of Dualism, i.e., there are totally separate "subjects" and "objects". For the most part, eastern world culture (and most any indigenous society anywhere in the world) accepts the concept of Holism. Holistic societies adhere to the idea of the deep interconnections of all things living and nonliving. Our planet is a single integrated system consisting of four deeply interconnected spheres: the atmosphere, hydrosphere, geosphere (aka, lithosphere), and biosphere. Basically, holistic people believe all things are not only connected, but also "sacred". In short, those people essentially are ecoethical. In general and overall, dualistic people are not. [I imagine there are exceptions to that statement.]

Given that, it's fairly easy to guess which of the two broad cultures would attain success in the sustainability arena. Unfortunately, it's the dualistic worldview which dominates the world today. Along with separate subjects and objects, there have arisen the ideas of possession, accumulation, and control of objects...in the extreme. Many of those objects are natural resources. Such objects are viewed as having only utilitarian value, and no intrinsic value. That's the antithesis of ecoethics.

So, specifically, what do we mean by ecoethics? Here's a brief summary of the key components.

The entire Web of Life is not only valuable, but to be revered (so to speak) as well. It's a deeply interconnected sphere (the biosphere), and every connection is important to the functioning of the whole. For example, consider bacteria for a moment. Most of them are either beneficial to human life and all other life, or harmless. Only about three to five percent are pathogenic to living humans. Both within and on the surface of your body, your physical self is teeming with beneficial bacteria. In general, without bacteria (and fungi), all plant and animal bodies would have to be burned after death, or disposed of in other ways. Nutrients for new organisms would be lost. Decomposers are crucial to operation of the nutrient cycle, and thus to the biosphere. If all bacteria became nonexistent, life on Earth would not survive. This is but one example of "lowly" organisms being critical to the survival of all life on Earth.

Ecoethics is the antithesis of the prevailing view of many which contends that Earth exists primarily (or only) for the use of humans. Not only the Earth, but all nearby solid cosmic bodies (our Moon, Mars, etc.) are subject to being dominated by *Homo sapiens*. Another word for such a view is anthropocentrism, the idea that humans are superior to all other forms of life. In general and most likely, such a view springs primarily from the Abrahamic religions – Judaism, Christianity, and Islam. While those religions (not including their extremist elements) do contain many worthwhile concepts – compassion for others, ethical conduct, love of family, etc. - they also have exaggerated the "superiority" of humans. At least one exception to that is the Society of Friends – The Quakers. From all I can determine and in general, that group appears to have a genuine ecocentric worldview. As well as for the abiotic environment, they have a deep respect for all living things and a bonafide interest in protecting the ecosphere.

Springing from the Deep Ecology Movement, ecocentrism recognizes the ascendancy and advantages of humans, but contends that such does not mean we are separate from and superior to the entire interconnected Web of Life. Rather, <u>we are part of it and dependent upon it</u>. *The ecocentric view is not proposing* (as some people believe) that all life forms are equal, but instead, all of them are deeply interconnected and dependent upon each other. This means ecosystems must be treated with respect and care. Some say, they should be treated with love.

Ecoethics, if it isn't implemented in a totalitarian manner, promotes the well-being of almost all organisms (including humans), sustainability on Earth, and a bright future for the Web of Life. Nevertheless, Charles S. Brown has pointed out in *Anthropocentrism and Ecocentrism: the Quest for a New Worldview* (see References), both views have value. Especially at the extremes, either one by itself can lead to massive ethical and biophysical problems. What's needed is a balance between the two. Currently, we're nowhere near that point. Anthropocentrism rules. That can change through education, discussion, and contemplation.

There's an old saying: if you want to change the world for the better, change yourself first.

Chapter 11: Solutions, and The Future

Micro and Macro Levels

Some solutions to the Crisis can be implemented (or at least promoted) on a personal/household and/or local scale. That's the "micro" level. Others – the "macro" level – must be instituted by States, Nations, and global organizations. Even at the macro scale, though, we as individuals can lobby (so to speak) for the implementation of various solutions. Whenever possible, we should. [See "Appendix III".]

Politicians seem to love inertia. They are often busy evaluating sustainability issues, but little of real significance seems to change. Peaceful pressure from the bottom up is required. Lots of it. That's how genuine change happens – through persistence over time. Once in a blue moon, it occurs due to a catastrophic event. Let's not wait for that. We're already having devastating events; but they're almost a picnic compared to what's coming *within* the next ten years or so if we don't act NOW.

Small is Beautiful

The title of this subsection is from the brilliant classic, *Small Is Beautiful – Economics as if People Mattered* (1973), by E. F. Schumacher. [See "References".] Not only is the book still relevant today, but one easily could argue that it's even more relevant than it was back then. It truly is a common sense economic treatise for the Ages.

In today's world, if we look at different scales of economic activity, in general a few conclusions strongly suggest themselves. First, the larger the scale (whether it's in relation to a single production facility or some aspect of the global economy), the more at risk is a worker's psychological and "spiritual" well-being. Large operations do not concern themselves much with the worker. Instead, the goods produced are the primary focus. If automation can replace human workers, all the better. If humans must be retained, then division of labor is implemented in most cases. Each worker completes a tiny portion of the overall job, and does so over and over and over. It's mind-numbing. It tends to wound one's spirit.

The second conclusion is: the larger the scale, the poorer the quality of the goods produced. Even if the product has a degree of quality/durability, generally it will have more flaws than one manufactured in a smaller, less mechanized facility. The third conclusion: if the geographical scale relative to distance from manufacturing to consumers is large, then the supply chain will be much more fragile. We've certainly been experiencing that lately.

Bottom line: planners and policy makers must shift to smaller economic scales if we are to improve the well-being of humans. Main Street should be working at convincing the Powers-That-Be to think locally and regionally rather than globally in terms of the production and distribution of goods. Using the USA as an example, the primary reasons mega manufacturers have moved production facilities out of this country are: dirt cheap labor, avoidance of environmental/health/safety regulations (responsibility), and a lack of concern for their fellow citizens. None of those are good reasons. None are ethical or demonstrate compassion.

So, if you're not a public or private policy maker, what can someone on Main Street do to help solve the too-big-to-worry-about-well-being problem? We should support smaller local and regional manufacturers and farmers to the best of our abilities and circumstances. In effect, that's a boycott of goods coming from monster junk factories and industrial farms. Support your local furniture maker or other durable goods manufacturer, farmer's market, organic farmer, beekeeper, etc. Whenever possible, support co-ops. Buy locally or regionally. Plus, we should express interest in smaller economies whenever appropriate.   If they rarely ever hear or read about small-is-beautiful, policy makers will continue to go down the same disastrous, vacant, economic path which has produced workers unfamiliar with the joy of work.

None of the above is to say that international trade has no place in the world. It does. The problem arises when such trade makes up almost all of a country's economic foundation. That scenario makes a nation less resilient to economic "black swan" events, and causes workers to be more vulnerable to job insecurity. In addition, the environment suffers damage from fuel combustion necessary for massive, long distance transportation of goods, and from manufacturing in areas which have few environmental regulations.

Universal Property

In 2001, Peter Barnes (see "References") came up with a brilliant idea: universal property, and how to use it. It sounds much like "public" property, but it's more than that. The problem with public property is that the government sometimes acts as if the government institution owns it (& the revenues from it), rather than everyone.

Barnes' concept could contribute significantly to reducing anthropocentric greenhouse gases, AND could also reduce poverty while giving more than half the population of the world some small degree of economic security. **Implementation of it cuts across political party lines and ideological divisions.**

Similar to the dividends Alaskans receive from oil profits, the idea is to charge mega polluters of the atmosphere a fee and then disburse those monies *directly* to citizens as a universal basic income. Some, perhaps many, Oligarchs probably oppose the idea. Most likely, that's because they view air, most water, and a good deal of land as waste sinks which are theirs to use *at no cost.* Some Main Street people believe the same thing. That view is not reasonable, and must be opposed.

Paying for the use of universal property and distributing the proceeds to **everyone** is a significant idea, and makes perfect sense. It has been twenty-one years since the idea was proposed, and politicians have done nothing about it. We should revive the concept and push for its implementation.

Carbon and Climate

Life on Earth is carbon-based, and the Carbon Cycle moves that chemical element through the atmosphere, bodies of water (especially the ocean), rock and soil, and living things. Some of the movements take millennia to complete (as in the case of carbon in marine sediments or sedimentary rock on land). Other types of carbon movements occur annually, or in much shorter time periods (example: every time you exhale, carbon dioxide is released to the atmosphere).

At any one time and on a worldwide basis, the reservoirs of carbon (such as air, water, living things, and soil) each contain *billions of tons* of the element. For instance, the atmosphere can have up to 750 billion tons of carbon dioxide in it. Soil can store up to 1,580 billion tons at any one time. Movement of carbon in the cycle is accomplished by various means, such as volcanic eruptions, wildfires, the decomposition of dead organisms, the weathering of rock, the thawing of permafrost, the burning of fossil fuels, the plowing of soil, the capturing of CO_2 by green plants during photosynthesis, etc. It's a natural cycle, and Nature knows best.

The point being: carbon is necessary to life, and there's plenty of it on this beautiful planet of ours. Carbon is not the problem. Human activity ***overloading the atmosphere*** with carbon dioxide and methane (never mind nitrogen oxides, which result from burning fossil fuels and are more potent than CO_2) **is the problem**. So the solution should be obvious: as much as possible, stop using the atmosphere as an industrial and commercial toilet. We all know this, right? I believe we do, so the question arises: why do the Powers-That-Be keep postponing the transition from fossil fuels to cleaner, greener options?

Most likely, the answer to that question is complex and multi-pronged. Nevertheless, I believe a major portion of the answer is fairly simple. Most politicians in DC are either neoliberals or puppets of same, and neoliberals are hanging on to oil with a death grip. That means Main Street will have to "lobby" those politicians in a dedicated, persistent manner in order to keep them from kicking the can down the road. If that doesn't do the job, vote them out of office. It's not flashy work, but it must be done. [See Appendix III.]

Biodiversity

<u>Here's one thing every one of us can do to help alleviate biodiversity loss</u> (as well as help remove CO2 from the atmosphere). Join the nonprofit Arbor Day Foundation. For about ten dollars or less from anyone who joins, they will have a few trees planted in damaged forests. Thus far, they've planted literally 100 million trees in America's public forests in just this way. They also have beautiful gift cards which can be purchased, and inside each one is written (paraphrased): in your honor a tree is being planted in a national forest. A few trees may not sound like much, but when thousands and thousands of people get involved, it adds up quickly to a significant conservation effort. When you join the outfit, they'll send you ten free tree seedlings/saplings which are appropriate for the area in which you live. Or, if you prefer, they'll have those trees planted in a forest of their choosing.

The Foundation has existed since 1972. Currently, they have over one million members, donors, and government & corporate partners. In addition to planting millions of trees in the USA, since 2017 the Foundation has been working in twenty-six other countries.

In 2019 alone, they facilitated the planting of thirty-two million trees in forests and towns around the world. This is a real boon to stemming the tide of global biodiversity loss. [See arborday.org/about/] It also helps reduce the overload of atmospheric CO_2. Spread the word.

Not only forests, but all other types of natural habitats are necessary for the fostering and protection of biodiversity. As much as possible, we must oppose all efforts by those who desire to open wildlife refuges and other public lands to commercial development. Any other unnecessary ecosystem damage/habitat destruction also should be opposed.

Our Toxic Environment

DDT (dichloro-diphenyl-trichloroethane) was the first modern artificial/synthetic pesticide. It was developed for large-scale use in the 1940's, used successfully in WW II to kill mosquitoes and lice, and by the 1950's was commercially available to consumers as the ultimate insecticide. Farmers used tons of it over the years to protect their crops.

Because of DDT use during the 1950's, the prevailing attitude was that humans finally would be able to control Nature relative to crop destruction and the spreading of disease by insects. Six thousand different synthetic pesticides would save us all. Chemists who developed them believed they were serving humankind in the highest manner. Very few safety studies were conducted on any of the hundreds of different poisons which were produced during that time. Those done usually dealt only with the person doing the spraying or dusting of the chemical. For the most part, no thought at all was given to effects on the rest of us, or on wildlife, or on beneficial insects (e.g., pollinators), or on water habitats, or on beneficial & necessary soil microbes.

In the late 1950's and early 1960's, biologists & ecologists began to suspect all was not well with the chemical utopia. In 1961 and 1962, studies determined that DDT was causing eggshell-thinning in pelicans and raptors. Populations of those birds crashed due to eggs breaking before embryos were developed into viable chicks. The parent birds crushed their eggs while brooding them. At the same time, insects began developing resistance to DDT. Mutations resulted in changes which enabled the mutants to circumvent the neurotoxic effects of the chemical. Some farmers reacted by applying even more DDT to their crops.

In 1962, Rachel Carson's book, *Silent Spring*, was published. She was not against the use of pesticides, but rather, their *overuse and indiscriminate use* . She also stressed the need for comprehensive safety studies relative to persistent, synthetic chemicals and their long-term effects. Her book was the beginning of the end for DDT. The chemical finally was banned in the USA in 1972. Not all countries did the same. DDT is still around today.

Since 1972, a few lab studies have indicated a possible link between DDT and breast cancer. It's generally classified today as a probable carcinogen. Most all people of my generation have DDT deposited in some of their fat cells. [To a somewhat lesser degree, I imagine subsequent generations are in the same boat.] Back in the day, that chemical was found everywhere – in soil, water, air, food, houses, yards, you name it. [Picture it in your mind.] Despite that, no long-term studies yet have been funded to determine the deleterious effects of DDT and its breakdown product, DDE, on human beings.

By 1976, another now ubiquitous poison was commercially available. It's an herbicide and is marketed under the brand name, "Roundup". The main ingredient is the chemical, "glyphosate". It's touted by industry as both safe and effective. It may be effective, but it's extremely doubtful that glyphosate is safe. In 2015, the World Health Organization (WHO) declared it to be a "probable carcinogen".

Silent Spring essentially launched the modern day Environmental Movement. The first Earth Day was in 1970. After that, the 1970's saw a plethora of environmental protection laws passed by Congress. A few existed prior to that time, but nothing compared to the '70s and even the early '80s. Then something burst upon the economic scene in the so-called Free World. Though it was nameless in the general public, it was in fact, neoliberalism. See Appendix I.

The philosophical underpinnings for neoliberalism were laid academically decades before the 1980's. But then in the late '70s and throughout the '80s, '90s, and beyond, its practical application was implemented. Today it's the dominant force in the world of economics. One might ask, so what?

Here's only one egregious example of corporatism at work during that period.

Decades and decades ago, DuPont Chemical introduced *teflon,* the non-stick coating for cookware. The chemicals (especially PFOA) in teflon also became incorporated into many other consumer items, such as water resistant fabric, stain resistant carpet, dental floss, microwave popcorn bags, fast food wrappers, and others. DuPont knew and kept a crucial secret from consumers regarding those chemical ingredients: they are highly toxic, and eventually became linked to six different diseases, including kidney cancer. For years, DuPont discharged wastes containing those chemicals into the waterways adjacent to their manufacturing facilities in Parkersburg, West Virginia.

In the 1980's and 1990's, people began to notice health problems such as cancer and liver disease in the area. There was a question, too, of birth defects. Plus, over 150 cattle died of some strange, blood hemorrhaging condition. Suspicions were high that DuPont manufacturing activities might be linked to those health problems. Some residents of Parkersburg, former DuPont employees, a cattle farmer, and lawyers launched what turned out to be a seventeen-year effort to find out the truth of the matter. The main attorney, Rob Bilott, was especially instrumental in bringing a class action lawsuit against DuPont.

In addition to discharging waste into waterways, DuPont buried seven thousand tons of contaminated waste in a nearby landfill. Later, it was discovered that the teflon chemicals were in the drinking water of the region.

It's estimated that 99% of Americans have these chemicals in our bodies, and most of the population of the rest of the world as well. That's because the waste also was discharged to the air until 2013; DuPont finally stopped manufacturing the substances during that year.

For the full story, see the *Democracy Now* news broadcast for January 23, 2018, democracynow.org

In my opinion, neoliberal policies and practices (both public and private) have set back by decades both environmental protection and the entire Environmental Movement. Consider the following. 1) Chemical product labeling laws either have been rolled back or circumvented with claims of "proprietary formula". 2) In 1980, Congress passed the Comprehensive Environmental Response, Compensation, & Liability Act (CERCLA), commonly known as "Superfund". Superfund sites are the worst of the contaminated hazardous waste sites in our nation. After **40 years of cleanup, we still have about 1,335 sites** – scattered across every State – which are grossly contaminated. **Only about 425 sites have been decontaminated since 1980.** As of April 2020, **51 new sites have been proposed** as additions to the National Priorities List (sites for cleanup). Apparently, the whole thing is no longer a priority in the minds of politicians, bureaucrats, and the public. 3) The Endangered Species Act (ESA), a bipartisan effort, was passed in 1973. It has been fairly popular with the American public. Since the 1980's, various parts of it have been gutted by legislators and/or the Executive Branch. The latest gutting was by the lameduck Trump Adminstration in December 2020.

Perhaps more important than any of the above are international Free Trade Agreements, which are spawned by neoliberals and enforced by the World Trade Organization (WTO). These Agreements provide private "courts" in which a corporation can bring suit or an arbitration case against any government that causes damage (including loss of anticipated profits) to their operation or property by passing legislation...such as environmental laws. Yes, you heard it right. A private "court" (run by the WTO and made up of corporate members) essentially can decide to override democracy. If the government does not comply with a ruling, the WTO can reduce Trade to that country. This really stifles the efforts geared toward environmental protection. Legislators perhaps are less likely to pass laws which ultimately cut into the profits of transnational corporations. It's not only environmental laws that are impacted; so, too, are health & safety bills – OSHA-type legislation. Such are the results of neoliberalism.

When it comes to solutions in relation to our toxic environment, this is a macro-level issue. Governments must bear the brunt of overcoming this challenge. Nevertheless, the rest of us have a role to play, too. We must continue to: use the many toxics found in our home with great care, and properly dispose of toxic chemical containers; always look for alternatives to toxic substances; *advocate for tough toxic substances laws; and advocate for Superfund site cleanup.* Finally, whenever possible BOYCOTT the use of toxics simply by *not buying them.* Corporatists are keen on letting the Market solve our problems, but we consumers haven't taken advantage of that to the fullest extent.

Nuclear Idiocy

Nuclear weapons are the most dangerous anachronism on Earth today. Some high-ranking military personnel have stated nukes are completely unnecessary for the defense of any nation. Despite those statements, and the fact that accidents involving weapons easily could result in nuclear war, and the fact that even so-called "tactical" nukes would be, if ever used, an ecological disaster, these obscene instruments of death continue to be developed and stored in several countries.

Politicians in those countries apparently are stuck in the Past. That must change if we are to live in a sane world. It's somewhat stunning that the citizens of the globe have not overwhelmingly demanded it. Thanks largely to ICAN, some progress toward that goal has been made (see Chapter 5). If nothing else, all of us can do one simple thing to help – become a member of ICAN.

The Plastic Problem

It has been estimated that of all the plastic ever produced, *only about nine percent* has been recycled. Despite that, the plastics industry has used the recycling tale to encourage consumers to buy more and more of their products. Other disturbing facts include: microplastics have been detected in our bodies; in parts of the USA's West, clouds of microplastics have "rained" down onto the land; and, these ubiquitous particles are now distributed worldwide in our food, fresh water, the ocean, on agricultural land, in our atmosphere, and in numerous species of animals, including livestock… and in us. Even clothes now are being made of plastics, and that contributes more microplastics to the environment via "wear and tear" of garments. In addition, when such clothes are laundered, household tumble dryers now spew microplastics (and smaller nanoplastics) into the atmosphere.

Unfortunately, studies regarding the effects of microplastics on living tissue currently are few and far between. We do know that "plasticizers" added during the production process of plastic contain various toxic substances. That alone should be a loud call to get a handle on this problem. But, what can be done?

Common sense would suggest at least three things should be undertaken immediately. First, we all need to cut back on our purchase and use of plastic. Boycott the production of it by not buying it. The monumental mountain of plastic junk produced worldwide every day of the work week surely must be staggering. Much of it is unnecessary. When I was a pre-teen kid, there was *no modern-day plastic at all.* It hadn't been invented yet. We did have a hard, brittle substance similar to one type of current plastic. It was made from formaldehyde and phenol (not petroleum), was called Bakelite, and was the first synthetic plastic. It was invented in 1909, and used primarily in industrial and commercial applications. Unlike the little kids of today, when I was about five or six years old all my toys were made of wood, metal, or cloth. That changed in the early 1950's.

The second thing which must be done is to invent a truly biodegradable plastic. Some progress has been made with plant-based "bioplastic", but it's not always 100% digested by microbes in field (rather than lab) conditions. Then, too, there's a problem with the petrochemical industry's development of "oxo-degradable" plastic. Like most plastic today, it's made from petroleum. Oxidizer additives supposedly make it break down much, much faster than normal plastic; however, so far it appears as though all that does is create *microplastics*, which are persistent in the environment. They are not digested by microbes.

The third thing which can and should be done immediately is to ban all plastic grocery (and other consumer goods) bags. There's no good reason why shoppers can't bring their own cloth bags to the store. Such bags last for years, and are relatively inexpensive to buy. Store management often doesn't like empty bags brought into an establishment, but they'll be able to adjust. Most grocery stores don't mind it at all. Finally, in the same vein, single-use plastic bottles also should be banned. They're ubiquitous only because production of them is much cheaper than glass, and they don't break (most of the time) if dropped. Those benefits are not worth the problems they create. Admittedly, such a contention is arguable; however, I believe most informed people would agree.

With the slow-moving shift away from fossil fuels finally in motion, Big Oil can see the handwriting on the wall. Movers and Shakers in that industry have a Plan B, and it has started already: drastically ramp up the production and marketing of *single-use plastic*, from water bottles to food trays to shipping box cushioning (bubble wrap, plastic "pillows", etc.). Plan B started because of less driving due to Covid. Single-use plastic now constitutes about forty percent of all plastic used.

Almost everything about the recycling of plastic has been misleading. For instance, the recycling symbol (a triangle of arrows chasing each other) found on plastic containers *does not necessarily mean* the container will be recycled. There are many different types of plastic, and about half of them either can't be recycled due to their specific chemical makeup, or won't be recycled because it's too expensive tondo so for that particular type of plastic.

The Social Issues

Including inequality and inequity, many various and egregious social issues are contained in today's ongoing Crisis. Due to the intentional brevity of this book, it's not possible to delve *in depth* into any one of them here. A few were touched upon in Chapter 1. Nonetheless, following is my hypothesis as to the primary cause of almost all of these social injustices, and what is necessary to mitigate the problems.

Ever since the beginning of *Homo sapiens*, we have adhered to the concept of tribalism. Even today, it's a way of life among humans. It's a philosophy and practice which more often than not provides solidarity, cooperation within the "tribe"/group, security, purpose, friendship, shared values, and stability. Modern-day tribes are numerous and varied. They are found in the arenas of religion, Race, ethnicity, economics, philosophy, nationhood, regionalism, vocation, academics, science, politics, ideology, and on & on. Often they overlap to one degree or another. A person easily can belong to more than one tribe. It appears, though, usually there's one which is the most influential in a person's life, and that can vary substantially from person to person. With any one person, it also can vary during different stages of an individual's life. Unfortunately, *there's also a dark side to tribalism.*

Hypothesis: one of the main causes of our social ills is the dark side of tribalism – intolerance of other groups, and arrogance relative to the virtues of one's own group. Especially when members of any "tribe" feel threatened, those aspects spring forth. It may result from a perception or misperception that the group's *culture* is slipping away or being assailed by some other faction. Prejudice follows, and discrimination. Social disharmony follows. Uncivil and unethical behavior follow. Sometimes, violence results. Intolerance and arrogance disrupt society in innumerable ways.

What can be done about it? What's the solution to the problem? For centuries, probably for a few millennia or more, enlightened people have been trying to promote this idea: the most important human group we all belong to is *Homo sapiens*. [Not only that, but the most important planetary group we're in is the *Web of Life*.] Relative to their efforts, some progress has been made... but not enough. One even could argue that for the last fifty years or so, we've been regressing.

For us to survive and thrive as a species, every one of us should look deep within and make a herculean effort to overcome intolerance. Such would involve countering the indoctrination we may have received from parents, other family members, friends, groups, institutions, and the like which has shaped our worldview. Entertain this probability, or at least, this possibility: in general, we humans have become a massive number of specialists in various ways. Apparently, to some degree we've lost the ability to think *comprehensively* about each other and our habitat. That seems to have resulted in many people developing a fear of losing their way of life, their culture. Intolerance grows in such a scenario, perhaps even hate.

For all their good works and best intentions, even religions at times exacerbate the situation. All across the spectrum, political groups *often* do the same. To be an ethical human, one has to look deeper than the surface of institutions and their ideologies. There's a fine line between "truth" and opinion. Even some theoretical physicists have stated: we humans barely have scratched the surface of Reality. We don't know nearly as much as we think we do. Question everything.

Bottom line: intolerance and arrogance increase the chances of a miserable human future, and ultimately, human extinction. That's a well-founded opinion. In your own life, be rid of the dark side of tribalism. Then humans and the entire Web of Life have a great chance to flourish.

Energy

Ongoing research with the natural mineral, perovskite, looks promising in relation to increasing the efficiency of solar panels. New research involving storage batteries also is yielding good results. Instead of updating our nuclear weapons arsenal, our Fed Government should be doing more to subsidize solar improvements. *Do what you can to "lobby" for such action.* Remember, though, solar technology alone will not solve the energy problem.

Relative to biofuels, algae offer the best hope. Many varieties of them build biomass thirty times faster than crops such as corn. A few can double their biomass in six to ten hours. A big problem currently is the cost of converting algae into fuel. The Pacific Northwest National Laboratory is working to solve that problem. They also are working on waste (manure, wastewater, & waste food)-to-energy projects, as well as hydrogen fuel cell projects. See https://www.pnnl.gov/microalgal-research-pnnl-sequim on the internet. It's an informative site.

As electric recharging stations become more numerous and improvements in vehicle batteries take place, we should see an increase in hybrid electric cars and fully electric vehicles. Technical hurdles remain in that arena, and more so with the development of hydrogen fuel cells for vehicles, but the future looks hopeful.

It would look even more hopeful if we would redesign towns and cities in ways which would reduce the need for individual, engine-powered cars. Some urban sections could be reserved for only mass transit vehicles and bicycles/adult tricycles. Other areas could be blocked off to everything except pedestrians. The Netherlands serves as an example of progress in this mode of living. For one thing, you see a lot of bicycles there. [Urban motorized vehicle traffic congestion is an affront to civilized human existence.]

One last point regarding energy: where is the overall Plan for the *transition away from fossil fuels*? If it exists, it doesn't appear to be widely circulated or easily available to the public. Each nation should have such a document. Better yet, the entire globe should have some means of coordinating efforts in this regard. Not having that surely is a testament to our inability to cooperate with each other in the face of an existential threat. We all are in dire need of politicians who do more than talk-the-talk.

Regenerative Agriculture

Factory animal agriculture and industrial raising of crops primarily via monoculture techniques both are egregious means of food production. They are an insult to us as supposedly enlightened beings, an insult to nonhuman life, and damaging to Nature in general. Both generate the overuse of pesticides and, in the case of factory meat production, inflict cruel harm on sentient beings. Both pollute the environment and are a burden on ecosystems. Factory farming is not only an ecological problem, but also an ethical problem. Higher animals often are egregiously mistreated.

Regenerative agriculture is a common sense alternative. It builds topsoil, which by the way, *is more valuable to humans than oil*. This type of agriculture is kinder to plants, animals, and ecosystems. That's especially true of the soil ecosystem, which is the basis of most all our food, and is an important carbon sink.

Wes Jackson and The Land Institute have created a revolution in sustainable, regenerative agriculture. They've been at it since 1976. The focus of much of their work is on perennial crops from annual crops.

Grains constitute about 70% of the calories we humans consume around the globe, and approximately 70% of the world's croplands. Most of those grains are annual plants. Raising annual crops means the soil has to be disturbed every year in order to plant, cultivate, and harvest the grain. Disturbing soil with plows, discs, cultivators, etc. results in carbon escaping into our atmosphere. Not only that, but soil erosion is orders of magnitude greater than it is on perennial crop fields.

Perennials don't have to be planted every year. The soil remains relatively unbroken, and the plants are more robust. The soil ecosystem continues to function as a carbon sink, rather than a carbon emitter. Erosion of soil is reduced. Fertility is enhanced.

Fresh Water Crisis

I'm not sure how many people *not* steeped in natural science are aware of the

 following, but suspect the number is relatively small. It's an important concept, and one crucial to understanding the fresh water crisis the globe is facing. Here it is---

A key component in the natural Water Cycle are "watersheds". Each one is an area of land which contains smaller streams and rivers that drain into a large river, lake, wetland, or some combination of those. A watershed often is referred to as a drainage basin. Such areas are separated by the ridges of hills, mountains, or gradual inclines. All land areas are made up of large and small watersheds. The term "land" relative to these areas includes all the vegetation growing there. Also, as well as surface water, groundwater is part of every watershed/drainage basin. Not only humans, but all other life as well need healthy watersheds to survive.

So, what's the problem? In a nutshell, our industrial societies and rat-race economies are significantly misusing and damaging watersheds. These areas comprise our natural water system from which we appropriate precious fresh water for our artificial water systems...IF the water has been held long enough for us to do so. If it isn't held long enough, the water drains to the salty ocean, sometimes relatively quickly. What holds it? **Vegetation**, living plant cover (especially trees), dead plant litter, a fertile and well structured soil all enable water to percolate into the ground, or to drain more slowly as surface water. Worldwide, we humans are harvesting too many trees (fifteen billion per year) and re-planting too few (five billion). We're paving over too much natural ground. In other ways, we're destroying too much plant cover of all kinds. It all contributes to diminishing *available* water supplies.

While urban and suburban water use is in significant amounts, **food production** accounts for **80%** of all water consumed in the United States. [That's according to a 2016 Report by the US Water Alliance, *One Water Roadmap: The Sustainable Management of Life's Most Essential Resource*.] "Production" in this case refers primarily to agriculture, but also to the manufacturing of processed food. Much of the field irrigation in farming is provided by groundwater aquifers. Those are being depleted at unsustainable rates. Large, "industrial" farms are mostly responsible. When it comes to water, bigger farms are not better. [The same is true when it comes to soil.] Smaller farms tend to use water more wisely, less wastefully.

Another aspect of the water crisis---

We Main Street people have made a grievous error by not strenuously objecting to the commodification and privatization of water and water services. Only governments ever should be involved in the allocation/distribution of public drinking water and irrigation water. Such resources forever should be in the "commons". Just like the atmosphere, they are universal property. Privatizing *public* water is an obscenity.

Until April 2021, three mega companies – Veolia, Suez, and RWE-Thames – controlled approximately 75% of the world's private water supply market. Then in April, Veolia and Suez merged in a $15 billion deal. Now only two companies control most of that market. Because the new outfit is French, and RWE-Thames is German, one might be tempted to think this is only a problem in Europe. That would be a mistake. They are all around the world... including in North America. Usually their company names are not prominent. Instead, they go by subsidiary names.

So, you may ask, what's the problem? Here it is in a nutshell... the reasons privatizing public water services is a mistake. [Rural cooperatives would be an exception.]

1. The price for water for the consumer almost always goes up, and significantly.

2. When the profit motive is involved and there's little to no competition, repairs and services often become spotty at best.

There's no one silver bullet which will solve our overall Crisis, and that includes the fresh water crisis. Following are some broad and hopefully obvious beginning steps re water.

1) Although our planet has a tremendous amount of water, only about <u>one percent</u> of it is available as fresh drinking water, or for cooking, bathing, watering lawns, crops, and livestock, etc. Each of us needs to *use it wisely*. When possible, cut back on usage.

2) Using the USA as an example, <u>it's time to end the Great American Lawn Fetish</u>. It wastes water, and artificial fertilizers & herbicides used on lawns run off, polluting streams & lakes. Lawn runoff contributes to eutrophication of bodies of water. Have you ever seen or smelled a eutrophic lake or pond? There are plenty of natural alternatives to grass lawns. So-called "beauty bark" is one; pea gravel is another; where appropriate, desert landscaping is another.

3) Eliminate leakage – not only in public water systems, but also in home and business water systems. According to the documentary, *Explained – World's Water Crisis*, undiscovered, small leaks are a big problem worldwide.

4) To anyone on farms and ranches of all sizes, *irrigate wisely*.

5) Lobby against privatizing public water systems. Maintenance of privately run systems can be poor at times, which results in more leaking pipes.

6) Do whatever you can to support smaller farms and organic farming. Large, industrial type farms pollute waters with nitrates and other chemicals by using excessive amounts of artificial fertilizers and herbicides.

The Food Crisis

Worldwide food production increased tremendously during the 20th century. A large part of that was due to the Green Revolution (previously described in the *"Soil"* section of Chapter 2). *The pace of increase has slowed considerably in this century*. Why? It's mostly because of the harmful, unintended consequences of that Revolution: the degradation of soil ecosystems due to overuse of artificial fertilizers and pesticides; and the depletion of groundwater aquifers due to poor irrigation practices in many areas. Other factors, such as more frequent droughts and destructive storms, play a part as well in reduced crop productivity.

The world leaders in the amount of food production (in descending order) currently are China, India, the USA, Brazil, and Russia. Overall, though, worldwide food production *and* distribution is not keeping pace with global population growth. The rate of population increase has slowed a bit compared to the 20th century, but today **about one billion people** still are undernourished. Even famine, or the imminent threat of it, is found in too many places. Part of the problem is supply chain fragility, while another part is most likely plain old inequity in the allocation of goods and land. But the probability is high that the main cause is human overpopulation. Contrary to some of the current propaganda, reducing human population can be done in a *completely* noncoercive and nonviolent manner. [See Chapter 9, "The Elephant in the Room".]

Ethics

I'm fairly certain that we humans all have been raised to believe human beings are superior to all other life forms on Earth. Perhaps not superior in every way possible, but at least in regard to intelligence and consciousness. As we journey through life, that idea often is reinforced by both people and circumstances. We view the concept as a given; no further thought about it is necessary.

I'm guessing some people believe in it because of religion. At least, that is likely in terms of the religions in Western World cultures. In those beliefs, humans are the pinnacle of creation. Other animal life, and certainly plants, are viewed as having only a utilitarian purpose vis-a-vis humanity. Forget about microbes. Atheists and perhaps agnostics, as well, seem to believe in it because of little more than Logic. It simply makes sense. Like most everyone else, for years I bought into the myth of human superiority. How could it not be true? Of course we were meant to dominate this planet. Eventually, though, I came to realize there are other ways of looking at the Web of Life and our place in it. I didn't experience a single "eureka" moment, nothing like that. It was more like hundreds of those moments, and they stretched out over decades. Because of my education and experience in Agriculture, Biology, and Environmental Geography, most eurekas involved nonhuman animals, plants, and other parts of the natural environment. Then, too, the ignorance (and sometimes, utter stupidity) displayed by we humans relative to our habitat and other life played a part in changing my worldview.

Here are only two little tidbits which helped to bring about about my transformation. **Did you know that trees communicate with each other in forests, and "mother trees" nurture nearby seedlings?** They do these things by means of symbiotic, micorrhizal fungi. Watch this video by the plant ecologist, Suzanne Simard. https://www.youtube.com/watch?v=Un2yBgIAxYs&list=WL&index=65&t=33s This is not based on magical thinking, but rather on solid ecological research. The clip is relatively short, and quite a revelation. The botanist, Diana Beresford-Kroeger, has something similar in a full-length documentary, *Call of the Forest: the Forgotten Wisdom of Trees*. Both pieces demonstrate scientific evidence supporting the hypothesis that trees have a type of consciousness and intelligence. Not like that of humans, but none-theless significant.

While we most certainly are unique, humans are but a single strand in the fabric of life on Earth. Our mistaken worldview coupled with our hubris regarding other life have re-sulted in the current Crisis. A change in our ethics is critical if we are to survive and prosper in the future. It has taken us decades to cause life-sustaining natural cycles to be on the brink of ruin. We must <u>now</u> re-double our efforts to avoid total disaster. That's not only my opinion. Many other people all across the political spectrum agree.

Politics and Sustainability

Two important factors---

1) Leaders around the world (with several previously mentioned and fairly notable ex-ceptions) seem to be essentially clueless when it comes to Ecology and Environmental Science. They appear to think Going Green means only solar panels, electric vehicles, gi-ant wind turbines, and recycling. A few probably add reuse and using less. Plus, it's not clear if they understand the current dependence of the manufacture and transportation of renewables on fossil fuels. On top of all that, they apparently know little-to-nothing about the importance of biodiversity in Nature, the extent of toxicity in our environ-ment, the nuclear waste problem, the nuke weapons problem, desertification, soil degradation, the extent of both income and wealth inequality and their impact, the problem of inequity, and on & on.

To compound the problem, the advisors of many Leaders are mostly *lawyers or main-stream economists*. All they seem to know is the status quo. On top of that, there's no (in the USA, at least) comprehensive, centralized Gov't agency with the purpose of implementing sustainability. I don't even know which current agencies are working on it, probably in isolation. None of the above bodes well for sustainability.

2) It seems 99.9% of the time when government begins working on some aspect of the socio-eco-econ-ethical Crisis we're facing, **too much emphasis is on the economy only.** Everything else is shoved to the back of the bus, or thrown under it.

....

For those who still believe government plus technology is going to solve the Crisis, I suggest this: **"lobby" hard for a national Department of Sustainability**. Sustainability has three major components: environmental, economic, and social. In such a department, *all three components must be equally represented*. It's a transdisciplinary field, and comprehensive thinking is paramount. This proposed department should be staffed with botanists, wildlife biologists, ecologists, agriculturalists, anthropologists, ecological economists, industrial ecologists, and perhaps an environmental geographer, a climatologist, and an ethics philosopher. [I imagine I left out an important field or two.] With all that, government *might* come up with significant solutions.

How will the new department be financed? There are numerous possibilities, none of which involve increasing taxes on wage earners. For example, tax the billionaires 0.1% on their incomes. They should be anxious to help finance the discovery of solutions. Why? **Because civil society is at stake**. Or, cut the Defense budget by 1% (or some such number). Or, cut the budgets of *every* current Department by some appropriate fraction of 1%. Or, believe it or not, there are many millionaires who are divesting themselves of a large portion of their wealth. Solicit their donations. Or, root out the financial waste in all departments and use that. Or, mix & match from the above. Or, just "print" the money. The Fed Reserve has been doing something similar to that for a few decades. It's not good for the value of the dollar, but then, *neither is the collapse of organized human existence.*

If national governments genuinely are serious about navigating through the Crisis we're now in, then a Department of Sustainability seems to me a necessity. At the very least, it would coordinate the efforts of other departments working on the numerous problems we face. History has shown that governmental departments often work in a manner which precludes comprehensive thinking. That has to change.

Chapter 12: Final Thoughts

Preface

An ancient Roman historian, Tacitus, wrote this regarding Rome's conquests: they make a desert, and call it Peace. Tweak it just a bit and such commentary now applies to modern-day industrial nations. Pursuing an expanding economy at any cost has resulted in desertification, biodiversity loss, diminishing forests, diminishing groundwater, a toxic environment, an atmosphere overloaded with greenhouse gases, social disruptions, and much more. They call it Progress. We have precious little time left to change that.

Our Greatest Challenge

It's not climate disruption, or biodiversity loss, or any of the other ecological catastrophes we face. **It's us**. We are our greatest challenge. Acting as though we're not part of Nature, we have caused the problems we're facing. Are we all equally responsible for the predicament in which we find ourselves? No, but all of us are facing the consequences and thus should be working together to stop the damage to our habitat and to our society.

As *Homo sapiens* attempts to move toward sustainability and all things "Green", it seems to me that we're still missing an important underpinning to the success of those efforts. It's known as ecocentrism.

In the dozens of articles on the circular economy, the doughnut/donut economy, etc. which I've read, the worldview of anthropocentrism appears to prevail. Is that, as the anthropologist Lynn White proposed decades ago, due to the Judeo-Christian worldview of the Nature-Human relationship (which sees humanity as both superior to and separate from Nature)? Perhaps, but I really don't know. It is true that historically Christianity, in particular, has claimed that only humans have a spirit. The Church believed (and most likely still believes) other beings in the animal world were/are devoid of such.

In any case and in general, humans do seem to be filled with hubris... especially regarding nonhuman life. That's a problem when we profess to be "going Green". In my view, there will be no successful sustainability without the adoption of ecoethics on a significant scale. Ecoethics will not be the prevailing philosophical standard without a paradigm shift in our view of the natural world. In the meantime, our efforts toward sustainability - **though laudable and a good first step** - *still amount to tinkering around the edges of the socio-eco-econ-ethical Crisis*.

Most Powers-That-Be seemingly are so ignorant of ecological principles, and so arrogant, that it's scary. That statement includes the entire political spectrum. Most of them have been doing their tinkering around the edges of the problem for decades. In a way, it would seem that Mother Nature has had enough of the arrogance of humanity. *Even the Green New Deal is short of what needs to be done.* A massive paradigm shift in lifestyle, politics, economics, industry, conservation, population growth, & ethics is required; but (in general) people in developed countries have been so spoiled for so long I fear that shift may not happen in time. Corporatism-Consumerism-Materialism might kill *organized* human existence either before or after we're over the tipping point. [It's not just corporatists who are at fault; it's also too many on Main Street.] There is some hope for a better outcome. It involves adopting the following worldview.

Nonhuman life has **intrinsic** value, not just utilitarian value. The Earth is not here for us to "subdue" (even though the Book of Genesis in the Bible urges humans to do so). We would do well to heed the wisdom of most all indigenous societies: though we're definitely unique, we humans are but one strand in the web of life. All sentient life is unique and "sacred".

In addition, as Barry Commoner stated in 1971 (in his book, *The Closing Circle*): "Everything is connected to everything else". We humans are not separate from Nature; we are Nature. We don't "come into the world" when we're born. We come from it. Essentially, that means all of Nature is to be revered.

All beings are Universal Consciousness expressing Itself. Why do I believe that? Admittedly, it's somewhat due to immersing myself in Buddhist scriptures (and writings about them), which started in 1974. In other words, it's a matter of faith in my choice of spirituality; however, there also are a number of experiences which strongly suggest it.

While doing field studies for various ecological projects (e.g., Environmental Impact Statements), it became clear to me that in any natural ecosystem – field, forest, wetland, etc. - all the producers, consumers, scavengers, and decomposers had an energy or spark of life which strongly suggested both a broadly common origin and a strong, *necessary* connection to one another. Nothing was separate and an island unto itself. Whether macro or micro life, all were interconnected, often symbiotic, and part of a synergistic existence.

I knew this from academic studies, but work in the field clinched it for me. It wasn't something immediately obvious, but after many projects over time, it became apparent. Such a thing – any part the web of life - is a beauty to behold. It causes one to become aware of the true value of Nature and humanity's place in it. We humans are grossly mistaken when we believe our species is separate from and superior to the natural world. We are deeply connected to the air, water, land, soil, plants, animals, and microbes of Nature. Our existence depends upon the whole.

Do we not breathe air? Are not water and food crucial to our survival? Do our bodies not house literally billions and billions of necessary bacteria? Do not green plants supply us with oxygen and food? Does not soil nurture land-dwelling plants which all serve us? By various means, do not animals supply nutrients to the soil, and do they not disperse plant seeds thus expanding a plant's habitat? Are not insect pollinators responsible for a significant portion of our food? Do not nutrients cycle through all living things, thus making decomposers crucial to all life? Obviously, YES, all of that and much more is true. I would think it's also obvious that humans (even mainstream economists) are inextricably a part of the natural world. We mistreat Nature at our own peril.

Without a holistic worldview, I don't see how we can live in harmony with the rest of Nature, or have any significant degree of sustainability. We must heed Barry Commoner's declaration: "Nature knows best". Yes, it's a paradigm shift in ethics, fraught with challenges; but it's necessary if we are to survive and thrive on Earth.

Many people around the world have made praiseworthy first steps toward transitioning to a sustainable human existence. For the most part, I think those efforts have been on a micro scale, and that's a necessity. Some of the efforts have been on a macro scale, and that's a good start. But too many Powers-That-Be still appear to be tinkering around the edges of the Crisis, or kicking the can down the road. For example, there's a lot of talk about "net-zero" (carbon emissions) by 2050. What that amounts to is postponing actions which could be taken now, and would result in solutions within ten years. Those solutions would not require new, unknown technology. 2050 will be too late; all the metrics strongly suggest such is true.

It isn't only carbon that's being dumped into the atmosphere by us. The natural Nitrogen Cycle is being overloaded with nitrogen oxides (NOx and N2O) by human activities - primarily factory farming, manufacturing, fossil fuel combustion, solid waste incineration, and wastewater management. Like methane (CH4), some nitrogen oxides are multiple times more potent a greenhouse gas than CO2. Currently, nitrogen oxide emissions from our actions amount to only about 7% of greenhouse gas emissions, but their potency is a concern. [NOTE: Don't confuse elemental nitrogen with nitrogen oxides. N2 gas makes up about 79% of our atmosphere. It's not the problem, and in fact, it's necessary to prevent the almost 21% O2 from igniting the air when someone lights a match. Oxides of nitrogen (which are chemical compounds rather than an element) can be the problem, and increasingly are something which must be taken seriously.]

Drastic changes are necessary to solve these problems, so all of us need to get on board and convince our Leaders to act, not just talk. We need to start phasing out fossil fuels NOW, *rather than issuing new drilling permits*. Similarly, we need to start transforming factory farms NOW. We need to support/encourage/subsidize family regenerative farms NOW, rather than providing welfare to fossil fuel corporations. We need to significantly reduce or eliminate the monstrous carbon footprint of cement/concrete manufacturing NOW. We need to tax mega carbon emissions NOW, and then disburse the funds directly as a universal basic income. We need to tax billionaires NOW. They stood on the shoulders of society to get where they are; there is no such thing as a self-made billionaire. Society provided them with all kinds of infrastructure, communication, education, contacts, law enforcement, a civil habitat, etc. [Yes, they helped pay for all that...but then, so did we all.] We need to start transitioning from an unlimited growth economy to a steady state economy NOW. We need to understand and promote the values of ecoethics NOW. Those are only a few of the things that could be done presently, but there's little to no action because of the undue influence of corporatism/neoliberalism/consumerism.

Synthesis

On the macro level, when we consider the deep interconnections among:

1) ecosystems/Nature and the many aspects of the eco-crisis, plus

2) the economy, finance, and rampant inequality, plus

3) mega business and corporate globalization, plus

4) politics, special interests, and propaganda, plus

5) industrial/factory type agriculture, overharvesting of fish and trees, monoculture tree plantations, plus

6) the ethical crisis relative to a lack of social equity, a lack of ecocentrism, and adherence to unlimited growth & overconsumption (never mind no government plan at all to totally eliminate nuclear weapons)...

We can conclude the following:

1) in order to survive and thrive, EVERYTHING must change, not just the energy sector;

2) the Establishment in government and industry/business is nowhere close to significantly and comprehensively addressing our multi-faceted Crisis; and

3) grassroots voices must be heard peacefully, no violence whatsoever, in order to bring about significant change. Gandhi, Martin Luther King, and others all had that right.

Sustainability

Environmental philosophers, public and private policy-makers, and others have been mulling over sustainability since at least the early 1970's. It's not something new. So, what is this concept, and why aren't we there yet?

The detailed definition of sustainability is not yet settled. It's an interdisciplinary field of knowledge, and so, many different people in many different vocations have significantly different views of it. Environmental philosophers, all manner of environmental scientists, public and private policy-makers, social scientists, economists, agricultural scientists, environmental activists, conservationists, and others all are involved in this field to one degree or another. Hence, until recently, the progress on the path to worldwide sustainability has been somewhat slow; some would say it has been agonizingly slow.

In brief and broad terms, sustainability means: **the ability/capability of humans to live and function in a manner which develops and preserves <u>the well-being of the environmental, economic, and social factors</u> of our existence for both the present and the future**. The challenges of such a philosophy are as follows.

Over the past few decades or so, disagreements have arisen as to *emphasis or focus* relative to the various aspects of sustainability. For example, some philosophers, policy-makers, etc. believe we need to prioritize the economic aspects by concentrating on "green economic *growth*". Anything else they tend to view as utopian, and unattainable. Others believe we should focus on environmental/ecological factors because the natural world sustains us with air, water, soil, forests, grasslands, food, building materials, minerals, fiber, and other resources. Still others believe our emphasis should be on social issues such as inequality, inequity, neocolonialism, neofeudalism, racism, injustice, and the like.

Another related example of the challenge of implementing a sustainable existence is what the political and environmental philosopher, Michel Bourban, refers to as "strong sustainability" v. "weak sustainability". [See Bourban in "References".] It's somewhat complicated, but a brief summary follows.

Both versions acknowledge and deal with the three main factors/means of production: artificial/physical capital such as infrastructure and machinery; human capital (workers and their abilities); and natural capital such as metals, lumber, and other natural resources. The difference between the "weak" and "strong" versions is in their emphasis on each of the means of production.

For the most part, mainstream economists and a large group of business owners adhere to the weak version. In that version, not much concern is given to natural capital, or ecosystems, or ecosystem services. Why? It's because the "weak" proponents claim in most cases artificial factors can be *substituted* for natural ones. *Substitutability* is the flavor of the day. For example, in many cases plastic can replace wood; it even can sometimes replace metal. Another example: petrochemicals such as nylon, polyester, and plastic fibers can replace natural clothing fibers (cotton, wool, hemp). This approach is referred to as "sustainable" because the champions of it throw in recycling, reuse, etc. to dress it up. There's more to it, but that's the gist of it. They maintain that the "weak" version is the only option. In their view, anything else is a pipe dream.

It doesn't take much thought to realize the focus of "weak sustainability" is strictly economic. There's little to no concern for the environment, social equity, or anything else other than the economy. Arguably (I suppose), it isn't really sustainability of any sort. Unfortunately, it appears to be very popular in the spheres of business, government, and academic neoliberal economics.

"Strong sustainability" is quite different. Theorists embracing this version are concerned with all the factors of our existence: environmental/ecological, economic, and social. They understand that all those factors are synergistically interconnected and the health of the entire system depends upon such.

Appendix I: Neoliberalism

As social beings, we humans recognize the importance of structure, organization, and institutions in everyday life. That's true all across the political spectrum. The amount of those pillars desired may vary from ideological group to group, but no one who is sane wants "Mad Max". So, we are reassured when we believe that politicians, religious leaders, economists, and others are keeping things in society humming along to the benefit of all. We feel secure to varying degrees, and life makes some sense.

The movers & shakers within those pillars know full well that it's crucial to maintain our faith in their abilities. Many/most of them have spent their professional lives working within and promoting a particular ideology or viewpoint. That's true whether the field is politics, or religion, or economics, or anything else. They truly believe in what they're doing and, for the betterment of society, would like to see it continue. [I'm excluding fraudsters and greedy sociopaths from the previous sentence.] Also, it's how they make their living.

The-Powers-That-Be and their apprentices learned this long ago: logic usually is not the best way to get people to adhere to your beliefs. Appeal to emotion is much more effective. Once emotion is generated, it's not especially difficult to transition to the final step: make belief in your viewpoint **faith-based**. Once there, any challenge to the system/belief/ideology is viewed as sacrilegious, and is not tolerated. Rather than engaging the "rebel" in productive and critical discourse, the heretic usually is attacked, deni-grated, and dismissed.

History

Once the philosophical underpinnings were laid in the 1930's at the U. of Chicago (and elsewhere), Professor Frank Knight and others began to spread the "gospel" of neoclas-sical/neoliberal economics. In 1947, the neoliberal think tank, the Mont Pelerin Society, was co-founded by Knight, Friedrich Hayek, Ludwig von Mises, Milton Friedman, and several others. Noted members have included many business moguls, and famous jour-nalists such as the liberal, Pulitzer Prize winner, Walter Lippmann, and the former radi-cal Max Eastman. Some of them rejected the term, "neoliberalism", but not the eco-nomic philosophy. They preferred the term, "neoclassical". They truly believed their new economics was going to spread freedom and lift the developing (or underdevel-oped) parts of the world out of poverty. The main themes were: free market econom-ics, free trade, and minimal (if any) government oversight. The market would regulate itself. [Note: years ago, I believed in all that. Over time, I came to realize that such poli-cies were resulting in egregious corporatism, inequality, and ecological destruction.]

It was quite an uphill battle for them. At the time, Keynesian economics reigned supreme. The neoliberals kept building their base. Meanwhile, the World Trade Organi-zation was being formed... which took years. After the socialist, Salvador Allende, was elected in Chile in 1970, Kissinger convinced Nixon that Chile was a "cancer" that would spread throughout the region. When the elected Allende was overthrown by a military coup (aided by our CIA) in Chile on 9-11-73, Pinochet was installed as the new leader of the country. Neoliberals convinced him to adopt their economic policies. Later, they considered the Chile "experiment" a success (relative to neoclassical/neoliberal economics).

In the economic disruptions of the 1970's, a few flaws in Keynesian economics became glaringly obvious. [Specifics are not relevant to the scope of our story here.] Then, with Reagan and Thatcher, came the Washington Consensus...

neoliberalism through & through. The neoliberals had made their first big breakthrough. Chile was small potatoes compared to Reaganomics. [I left out the Carter Administration, which was prior to Reagan. They did a bit of deregulation, and were a bit corporatist/globalist (mostly due to Brzezinski) but, like Chile with Pinochet, the Carter Crew were nothing in terms of neoliberalism compared to Reagan].

From then to present day, neoliberal/neoclassical/mainstream economics has been, and continues to be, sacrosanct. That's true even though covid-19 has seriously wounded it. For any economist (or really, anyone) challenging it, you essentially are guaranteed to be professionally suppressed and dismissed. There's no need for mainstreamers to defend their ideology because to them it's simply a given. They largely shift the blame for any problems--- inequality, ecological damage, increasing numbers of monopolies, etc.--- off their ideology and onto anything & everything else. Perhaps those problems are due to a combination of factors; nevertheless, it should be plain by now that neoliberalism is responsible for *at least* a large portion of them.

Flaws

The following flaws are ignored by mainstream/neoliberal economists---

1. Neoliberal economists seem to have little to no grasp of complexity theory. https://www.youtube.com/watch?v=hLXIJF5ytpM [The clip is only 6.5 minutes long, and gives an excellent overview of the concept.] Never mind (for a moment) their unrealistic belief that economics and economies are separate from, and larger than, Nature... which is bad enough. Perhaps worse is their view that the individual parts of any natural system are separate from each other. Thus, in their ignorant thinking, one part can be removed, damaged, or destroyed without any serious effects on other parts or the whole.

I can imagine certain economists saying to central planners: so what if we clearcut tropical forests? They're only trees. We need the lumber, and farmers/ranchers need the pasture or crop fields, or developers need the space for housing projects.

That kind of thinking allows the true believers in unlimited growth and overconsumption on a finite planet to pursue their shortsighted objectives. Any problems are considered negative "externalities", and not their concern.

2. In the same vein, they appear to fail to understand the creation of a "whole" that is greater than the sum of its parts--- synergy. The phenomenon is common in both natural and artificial systems. Ecosystems are without a doubt, synergistic, and they are our life support. For that reason, we must treat them with due consideration and care.

Decades ago and in a somewhat broader context, the esteemed macrohistorian, Professor Carroll Quigley of Georgetown U., gave the following assessment.

"Professor [Lynn] White's thesis is that when the Judeo-Christian faith established the view that there is no spirit in nature other than the human, the world was reduced to a created object to be exploited [conquered] by humans. Thus the way was opened to the destruction of nature and to the near total pollution of the world -- a consequence that may have become inevitable with the *rejection* [emphasis added] by the Church, in the latter thirteenth century, of the message of St. Francis to treat all nature as sacred.

The cognitive techniques derived from our underlying outlook have included (a) using analysis rather than synthesis in seeking answers to problems; (b) isolating problems and studying them in a vacuum instead of using an ecological approach; (c) using techniques based on quantification rather than on qualification study done in a contextual situation; (d) proceeding on the assumption of single-factor causation rather than pluralistic, ecological causation; and (e) basing decisions and actions on needs of the individual rather than needs of the group."
Quigley, Carroll, *Needed: A Revolution in Thinking*, National Education Association Journal, Volume 57, May **1968**, pp. 8-10.

Point being: warnings were given long ago. To this day, however, too many people seem to be ignoring them. While the revolution in thinking has progressed since the above article, we do need to pick up the pace. Immediately.

3. Mainstream economists and oligarchs have promulgated the following, and have indoctrinated Main Street to the hilt in this false narrative.

●Material wealth is always meritorious and indicative of success and happiness in life.

●Poverty is almost always the result of individual failing.

4. Whether it intended to or not, neoliberalism has installed massive central planning in the economic sphere... planning not by governments, but rather by central banks (which are privately owned), other mega banks, and other mega corporations. Governments largely go along with the whole scenario because politicians get campaign funds mostly from those private institutions. Don't bite the hand that finances you.

I don't believe I need to explain why that setup is a really big flaw in the secular religion of perpetual growth & overconsumption, or why it's damaging to Main Street. Such should be obvious.

Years ago, in a PBS News Hour interview during his time as Fed Chairman, Alan Greenspan stated: "We are completely independent. No one has control over us.". Given some of their testimony during congressional hearings after the 2008 Crash, mega bank CEOs apparently feel pretty much the same way.

Conclusion

If we are to thrive on Spaceship Earth, it seems obvious to me that we must address all the above in an ecological (holistic) manner. Except for a relatively few individuals and groups/institutions, we have been doing little more than tinkering around the edges of massive problems for decades. A major change is past due. Though they are relatively few, those who have been working for decades on solutions to our many, interrelated, and various dilemmas deserve our esteem, support, and thanks. Ironically, those people are too numerous to list. Plus, many are essentially unknown in the broader sense. Kudos to all.

The priests of the Unlimited Growth paradigm usually wind up extolling it via a reference to the glory of capitalism. But which capitalism are they talking about? Is it crony capitalism, or casino capitalism, or disaster capitalism, or the one in which a central bank uses central planning to totally eliminate a "free market" when it comes to currency/interest rates/other monetary issues? One thing is certain--- it's NOT the capitalism of Adam Smith. Not even close. Smith was for small business, not monopolies or mega business. Plus, why do those high priests ignore the fact that the USA and other democratic countries have had a Mixed Economy for decades & decades? That was true even during the Golden Age of American Capitalism (from the end of WW II to 1971).

In my opinion, there is no doubt that neoliberalism in politics, finance, and economics has been a major contributor to the following:
1) income inequality;
2) social inequality;
3) privatization and austerity;
4) ecological/environmental damage and destruction;
5) increased susceptibility to disease;
6) lost manufacturing;
7) a much less resilient economy; and
8) several other maladies.
Neoliberals have managed to rule the roost right through today primarily by the use of

massive Edward Bernays style propaganda. It saturates the public and private spheres every day. It promotes distorted ideas of both "freedom" and "capitalism".

It's time to make some noise... time to seriously consider Aldo Leopold's "Land Ethic"... time to stop being lemmings going over the cliff... time to get some ethical backbone... time to stop drinking the Edward Bernays style Propaganda Kool Aid... time to push for cooperation rather than division... for tolerance rather than hate... for clarity in politics rather than incoherence... and finally, it's time to promote with vigor the idea that this precious planet on which we live has limits which must be respected.

Appendix II: Mega Business

Preface

Arguably, one of the biggest threats to the ecosphere and to the implementation of sustainability is the artificial construct known as the *corporation*. Early corporations in the USA usually were chartered by the government to perform only one function, such as building a bridge or a road. Plus, corporate charters were good only for a limited time. Eventually, lobbying by business elites changed all that. Plus, the elites managed to get the corporation declared a "legal person", totally separate from its owners.

While individual corporate officers still can be charged with a crime, what happens more often is that the corporation is charged. If found guilty, because a corporation cannot go to prison it is instead ordered to pay a fine. In some cases, those fines are in the millions (or even billions) of dollars. In the case of a mega corporation, that amount seems to have little effect on future behavior.

Corporate Globalization – Early Players

Large transnational corporations (including mega banks) are responsible for significant parts of our current socio-eco-econ-ethical Crisis. That's largely due to corporate globalization. Here are the beginning seeds of the phenomenon known as Globalization.

In 1944-45, as a result of the *Bretton Woods Agreement*, the International Monetary

Fund (IMF) and what finally came to be known as the World Bank were created to help rebuild Europe after WW II. The *Agreement*, drawn up by the UK and the USA, established the rules/regulations for an international monetary system. Some years later, the IMF and the World Bank began making loans to poor countries, the Third World, and those loans had strings attached. More on that in a bit.

In 1965, General Suharto launched a successful coup against Sukarno in Indonesia, a country rich in resources, potential markets, and potential labor. The coup was known about beforehand by both the UK and the USA, but not revealed to Sukarno. [Sukarno, a strong nationalist, had kicked the IMF/World Bank out of his country...refused their "aid".] After the coup and over time, between 500,000 and one million Indonesians were murdered by the Suharto regime. Our CIA supplied a list to the Suharto regime which facilitated the beginning of the massacre of about 5,000 Indonesian communists. During that time, loans were made to Indonesia by the IMF and World Bank...with strings attached.

In 1967, Time-Life Corporation sponsored a meeting in Switzerland, the purpose of which was to plan the corporate takeover of Indonesia...that is to say, the economic (not political) takeover. David Rockefeller, a Council on Foreign Relations (CFR) member and representing Chase Manhattan Bank, attended along with dozens of other business luminaries - representatives of Lehman Brothers, various oil companies, food companies, ALCOA, U.S. Steel, Siemens, various other banks, etc. General Suharto had his representatives there as well. [This is a classic example of Fascism in action--- mega corporations colluding with government cronies in order to control both the People and the economy.]

The large group in Switzerland divided up according to various sectors: banking and finance, food, consumer goods, oil and energy, etc. Each smaller group then hammered out the rules for doing corporate business in Indonesia. As part of the deal, the government of that country obtained another series of loans from the IMF/World Bank.

In 1973, the USA's CIA helped overthrow the democratically elected President of Chile, Salvador Allende...because he was a Socialist, and probably had plans to nationalize U.S. businesses in Chile. A brutal dictator, General Pinochet, was installed; some time later, Pinochet's government obtained loans from the IMF/World Bank. Pinochet ruled for seventeen years. Henry Kissinger, a CFR member, chaired the meetings in the Nixon Adminstration that dealt with covert operations in Chile prior to the coup. Also prior to the coup, a Pepsi Cola franchise owner in Chile who also owned a newspaper (which was thought to be a CIA front) was called to DC to update Dr. Kissinger on the political situation.

In the 1980s, Thatcher of the UK built up her country's arms business and made a lot of armament sales to General Suharto of Indonesia. Also during that time, Thatcher and the USA's President Reagan laid the groundwork for what became known in 1989 as the **"Washington Consensus"**---a series of policies that supposedly would lead to global prosperity and unfettered economic growth by means of deregulation, privatization, and trade liberalization (so-called "Free Trade") around the world. In the Third World, such a lofty goal would be attained through use of the IMF and the World Bank. The "Structural Adjustment Programs" (now re-named) of those two institutions supposedly would be the key to success. Details later. This was all during the tail-end of and after the Pinochet regime in Chile, which the Neoliberals viewed as an economic success.

Corporate Globalization – More Players & the Methods

As you saw in the previous section, the IMF and World Bank are key players in the process of Globalization. [I consider them quasi-governmental bodies, rather than non-governmental.] Add to that the U.S. Treasury and Federal Reserve, as well as the central banks of most of the other industrialized countries. It is believed by many serious writers that the IMF/World Bank are merely extensions of the policies of the U.S. Treasury, the Fed Reserve, U.K.'s Treasury, and the U.K.'s Bank of England.

While the above institutions are important, they are no more important than the non-governmental bodies: the transnational mega corporations, the Council on Foreign Relations (CFR, founded in 1922) and its U.K. counterpart, the Royal Institute of International Affairs (aka, Chatham House), and the Trilateral Commission (founded in 1973 by David Rockefeller and Zbigniew Brzezinski). The CFR and Chatham House are the anglophile Round Table Groups, described in the late Professor Carroll Quigley's 1966 book, *Tragedy and Hope*. The best quote from the book: "*The tragedy is that we no longer have a representative government in this country, or any country; the hope is that the little people will come to accept that, because there is nothing they can do about it.*" Again...it was published in 1966. Quigley also claimed that the CFR was a front for the J.P. Morgan Company; keep in mind that the good professor (Bill Clinton's mentor at Georgetown University) was a very well respected historian and scholar.

In 1991, David Rockefeller stated publicly, "Surely it is preferable to have the world run by unelected, supranational, [business] elites rather than the nations of the past with their corrupt politicians." *That is the goal of Globalization*. The function of groups such as the CFR, Chatham House, and the Trilateral Commission is to provide the philosophical underpinnings and the unquestionable academic justifications, for the governmental policies which result in the implementation of Globalization. Consequently, the CFR publishes academic-level papers such as *Building a North American Community* (meaning, the North American Union), *The End of National Currency*, *Bretton Woods II: Does the World Need a New Monetary System?* (authored by Larry Summers, Obama's chief economic advisor), and *Global Economic Governance*. The Trilateral Commission pens scholarly papers such as *Global Governance*, *Global Redistribution of Power*, and *Sovereignty and Intervention*. The tremendous influence of these groups on governmental policy cannot be overstated; for all intents and purposes, essentially they formulate most of our nation's foreign policy and perhaps a good deal of our monetary policy.

The Methods---

There seems to be no particular chronological order to these methods; at times one must precede another, but mostly they are all taking place more or less simultaneously. I'll list them here, then expand on each one later: selection of Globalist candidates for national office, and then the marketing of them; introduction and justification of Globalist policies to the intellectual elite of the country; use of IMF/World Bank "Structural Adjustment Programs", thus facilitating U.S. corporate expansion to poor countries; Trade Agreements and the WTO (World Trade Organization); the use of "Economic Hit Men"; facilitating coups in uncooperative countries; the maintenance of approximately 800

U.S. military bases on foreign soil; and finally, invasion of uncooperative countries (which is a last resort).

Here's what a national election amounts to in the U.S.: the heads of transnational corporations and other members of the Oligarchy/ Corporatocracy, having identified and groomed pre-selected candidates from the two major parties, make massive campaign contributions (indirectly, through PACs) to both. One candidate or the other may be favored from time to time, but generally either one will do. [There is arguably only one major political party in the U.S.: the Transnational Mega Business Party, with a Republican faction and a Democrat faction.] As FDR once wrote to a friend, "In this country, candidates are selected, not elected." The campaign contributions are necessary because the candidates must be marketed to the voting public, and that requires expensive PR firms and multi-media advertising. PR firms run campaigns, not the DNC or RNC. The public usually is not bothered with any details regarding Globalization.

Globalist policies must be "sold" to the intellectual elites because those elites are extremely influential in the Corporate Media, educational institutions, various community groups, the entertainment industry, and government at all levels. A relatively small percentage of the American public actually pays attention to important public matters, and it's useful to have them supportive of Globalist tools, e.g., NAFTA (Por the revised version). As mentioned above, the intellectual think tanks (CFR, etc.) "justify" policies and tools to the intelligentsia via scholarly papers and even books. Often the mainstream Corporate Media present summaries of think tank ideas to the general public.

A key method in terms of the Third World is to convince poor countries to take on massive IMF/World Bank loans...for "development". Attached to each loan or interest reduction on a previous loan are conditions, Structural Adjustment Programs (SAPs), which require actions such as: cutting government expenditures, devaluing the currency, privatization of State-owned enterprises, lifting of trade restrictions, enhancing the rights of foreign investors, gearing the economy to exports, and removing price controls and/or State subsidies. Critics have referred to this as virtual extortion. As a consequence, Poverty Reduction Strategy Papers (PRSPs) have replaced SAPs; however, their contents more often than not turn out to be essentially the same as the old SAPs. The money loaned doesn't go to job development or farm aid, rather it ends up in the wallets of companies such as Bechtel, Hochtief, or Halliburton. They are hired to build ports (for those exports), roads (to get the goods to the ports), water systems, and other infrastructure.

The use of Trade Agreements (such as the original NAFTA) to implement Globalization requires no discussion, except to say there is some controversy regarding the status of NAFTA. Canada and Mexico claimed it was a Treaty; however, it received only 61 votes in the Senate. A treaty requires the approval of 67 Senators. Whether or not it's a Treaty is important because Treaties override U.S. laws. NAFTA was modified by the Trump Era, and as far as I can tell, it still contributes to the negative effects of Corporate Globalization – one of the most significant effects being the transfer of wealth to the Upper Crust of society.

The World Trade Organization (WTO) plays a major role in the implementation of Globalization. The WTO, headquartered in Geneva, Switzerland, began in 1995 (replacing GATT) and is the only international body dealing with trade between and among

nations. It makes the rules for 97% of all world trade, and publishes research papers such as, *The WTO and Direct Taxation*. No country can override the WTO's rules. Some critics have called the WTO an embryo world government, and claim that it favors rich countries over poor ones. In any case, it has impacted the sovereignty of member countries. A "treaty" of the WTO, the *General Agreement on Trade in Services* (GATS), extended the world trading system to include the service sector; all WTO members are signatories to GATS. The GATS Treaty is the primary tool of the WTO for the implementation of privatization of services such as municipal water delivery and delivery of

electricity.

Economic hit men (a la John Perkins, *Confessions of an Economic Hit Man*, 2004) are used to essentially extort/blackmail uncooperative countries, primarily poor countries. The "hit man" will point out to the leader of the country that his country has a World Bank/IMF loan that cannot possibly be repaid, thus the leader needs to do one or more of a number of things, such as voting a certain way in the U.N. on some issue, or favoring some private contractor regarding a large in-country project bidding process. "Oh, by the way, here's $100,000 for your trouble." Mr. Perkins was never 100% certain who his employer was, but eventually assumed that it was the CIA. His book was a *New York Times* bestseller.

Coups as a tool of Globalization should be self-explanatory. The example of Indonesia was given previously. Sukarno would not cooperate with the IMF, so he was booted out.

[Same thing with Allende in Chile.] The dictator Suharto in Indonesia allowed U.S. companies to move in and utilize virtual sweatshops for the manufacture of goods for stores such as the GAP and Old Navy. The U.S. and U.K. *have participated in coups in fifty countries since 1953*, when the CIA engineered the ousting of the democratically elected leader of Iran and installed the Shah. The Shah purchased millions of dollars of armaments from the U.S. and, of course, supplied us with a portion of our oil requirement.

The maintenance of about 800 U.S. military bases around the world is tied in to the final method: invasion of uncooperative countries as a last resort. A prime example is Operation Just Cause, the invasion of Panama in 1989. The dictator Noriega had become

uncooperative, especially relative to Nicaragua, thus his drug informant days for the CIA were used as an excuse to invade, without admitting that he was a CIA operative in the past. It all came out a few years later in a few books and the documentary film, *Deception in Panama*. At the time of the invasion, George Bush the elder was President here, Dick Cheney was the Secretary of Defense, and Colin Powell was the Chairman of the Joint Chiefs of Staff.

Other examples include our invasions of Afghanistan and Iraq. While 9-11 provided a fortuitous reason for the invasion of Afghanistan, Taliban representatives were told months before that date, "If you don't accept our carpet of gold [for a pipeline easement across Afghanistan], we'll give you a carpet of bombs." (According to a whistleblower) Plus, according to Brzezinski in 1998, control of Afghanistan, Iraq, Uzbekistan (where we now have a military base) and a number of other -stans in Central Asia was a "geostrategic imperative for American primacy". Read into that what you will. In the case of Iraq (which, even according to Dubya Bush, had nothing to do with 9-11), Saddam, our former ally, had become uncooperative relative to the petrodollar. Prior to the invasion, he was on the verge of accepting Euros for oil. If allowed to proceed, that likely would have ended the petrodollar and the dollar as the world's reserve currency. That, in turn, most likely would have resulted in a series of events ending with the collapse of our dollar. Plans to invade Iraq were formulated months before 9-11. Saddam, our big buddy in the '80s, had to go. [In the '80s, we supplied him with weapons in order to punish Iran for the overthrow of the Shah and the holding of American hostages. The result was the eight-year Iraq-Iran war.]

Corporate Globalization – Philosophical Foundations

In the Scientific Method of Inquiry, there is a very significant difference between a

hypothesis and a *theory*. The hypothesis is an educated guess, a proposed solution to a problem, and is generated after gathering information and making observations relevant to the problem. Prior to that, the first step is to identify the problem. A theory is a *tested* hypothesis with positive, repeatable results. It must stand the test of time before it is accepted as "fact". The Media often misuse the word "theory", as in their use of "conspiracy theory". My best guess is that 99% of all so-called "conspiracy theories" are, in fact, conspiracy hypotheses. A hypothesis is not "proven", even though it may contain some suggestive or even substantiating evidence. Any hypothesis can be based upon very little actual, conclusive evidence; a theory is a different matter.

I mention all of the above because what follows below is my hypothesis regarding Globalization. The problem I'm attempting to solve is: what caused Globalization, what brought about the ongoing process, what motivates the people involved? There's more to it than simply more profit, more power, and/or a more streamlined business atmosphere. By the way, I do NOT believe that it's a conspiracy. It's more along the lines of individuals and groups with similar interests and mutual goals working sometimes independently, sometimes cooperatively, to attain their objectives.

The first condition responsible for Globalization is the ongoing NeoImperialism of both the USA and the UK. We Americans like to think that our nation is not imperialistic. Scholars are divided on the question. Our history is suggestive of a definite imperialistic tendency, and we love the phrase, "Manifest Destiny". Consider these acquisitions of territory: Louisiana Purchase (1803, from France); Florida (1819, from Spain); California and the Southwest (1848, from Mexico); Alaska (1867, from Russia); Hawaii (1898, from natives); Cuba (temporarily), Guam, the Philippines (1898, from Spain); and, if you agree with Chalmers Johnson (former CIA analyst), our 800 military bases currently around the world may be considered neo-imperialistic "colonies" of a sort. All this is not even considering the conquering of American Indians. [According to Russell Means (formerly of AIM, the American Indian Movement) and John Echohawk (of NARF, the Native American Rights Fund) there is nothing whatsoever wrong with the term, "American Indian".] Some of the above territories were acquired peacefully, others by war. Finally, in 1898 Mark Twain founded the *Anti-Imperialist League,* mostly because of our involvement in the Spanish-American War. That seems to suggest that at least some people considered the USA to be imperialistic.

Along with our NeoImperialism, there is the concept of American Exceptionalism---the belief that we are more special, of greater good, than other nations. So, the argument goes, if we are in any way imperialistic, it is a benign Imperialism and for the good of the world. That concept is so indoctrinated/ingrained into our consciousness that few Americans ever question it. It is probably found in most of us, especially those who are the public and private Powers-That-Be. Included lately is the belief that we have some sort of Divine Right and Duty to "spread democracy" around the world...even if people do not want our help. [The majority of Afghan people wanted us to leave their country for a very long time; that same majority did not support their various corrupt governments.] This mindset partly is what gives rise to the "American Empire", a controversial term representing our 800 global military bases and our securing of access to foreign resources.

The next piece of the puzzle is Neoliberalism, which is an economic and social policy – or ideology - driven by privatization, deregulation, austerity, and trade liberalization on a global scale. [See Appendix I.] Most of the world, except for Americans, has been familiar with the term for at least twenty years. Supposedly, the first test of it as a hypothesis was in Pinochet's Chile in 1973, after the CIA overthrow of Allende; also supposedly, it was a resounding success...depending on how one defines success. It also is the basis for the *Washington Consensus*, which was brought about largely by Reagan and Thatcher. One might think Neoliberalism merely is some sort of re-packaging of Conservatism, but because it involves mega corporations working closely with national governments and international agencies, it is more akin to Fascism than anything else. Organizations (such as the WTO, IMF, etc.) made up of corporatists and international bankers are telling countries how to operate, rather than the reverse.

That brings us to Supranational Corporatism, involving the Crony Capitalists of transnational mega corporations and their government cohorts. Crony Capitalism is not genuine Capitalism. Instead, it is a collusion of CEOs and their cronies in the government to bring about less and less competition in the global, regional, and even local business arenas. Together they create a Fascist framework to support activities that virtually erase national trade borders, eliminate smaller competitors, and marginalize national governments. Because of the supranational corporatists, we are moving gradually toward a feudal-like society with only two classes---the super rich and the poor. In other words, we're moving toward neofeudalism. International bankers aid and abet the culprits by manipulating money and economies via central banks, creating economic bubbles and crashes that facilitate the looting of commoners. During the last crisis in America, the "Too-Big-To-Fail" companies became even larger, as in the cases of the bank mergers and buy-outs in the 2008 Crash.

The corporatists have been influenced in a most significant way by the philosophies of NeoImperialism, American Exceptionalism, and Neoliberalism. Globalization is a logical extension of "Manifest Destiny" to the mind of a corporatist, and who better to run the world than American elites, they ask themselves. It's also reassuring to them (probably on a subconscious/unconscious level) that whatever Americans do must be good for the world.

In the process, the corporatists have become haughty, deceitful, and disrespectful of We the People. Some of them refer to us as "the small people"; their arrogance is palpable. It's important to understand that the transnational corporatists live in an entirely separate, artificial universe. They appear to feel, in general, that they are way above the rest of us, that they can make their own rules and do pretty much whatever they please, legally or illegally. Cutting corners to save money is a given, even when the safety of people is involved. The past BP fiasco in the Gulf of Mexico is evidence of that---certain cheap materials were used in the well, and warning signs of disaster were ignored...not only by BP, but by Transocean and Halliburton as well. [By the way, British Petroleum was once a public/government operation; Margaret Thatcher privatized it, and it became the private corporation, BP. This is the same BP whose Head referred to the Gulf Coast residents as "the small people".]

The last piece of the puzzle is the Theory of Propaganda and Indoctrination, based on the work of Edward Bernays, the Father of Public Relations, and his most important book, *Propaganda* (1928). Bernays believed that propaganda (the manipulation of public opinion toward a desired end) was "an important element in democratic society". He went on, "Those who manipulate this unseen mechanism of society constitute an invisible government which is the true ruling power of our country." An important journal article of his was, "The Engineering of Consent", in which he equated propaganda with the article title. In the political arena, Bernays worked with the old United Fruit Company (Chiquita Brands International today) and the U.S. Government to bring about the overthrow of the democratically elected President of Guatemala in the 1950s. United Fruit then dominated a series of corrupt governments in that country. [A key element of the propaganda demonstrated that the President was a Communist, but he wasn't .]

We Americans seem to believe that other governments propagandize their citizens, but our government does not. I suggest that we are propagandized almost constantly by our government, especially regarding Globalization, the political party not in power at the time, and going to war. I recall a debate of sorts between Al Gore and Ross Perot in 1992 (maybe) on *the Larry King Show.* Gore was extolling the virtues and benefits of NAFTA...pure propaganda. Perot retorted with his famous (paraphrased), "I guarantee you, all you're going to hear is a giant sucking sound...the sound of jobs going to Mexico!". He was right; and we still have not disengaged our country from NAFTA (though it has been revised), which was approved in 1994. The poor people of Mexico are still poor. The only ones benefitting from NAFTA are the corporatists. Then, too, Congress approved CAFTA (Central American Free Trade Agreement) in July of 2005.

For several years, the Iraq War was a boon to a number of transnational corporations: Kellogg, Brown & Root; Blackwater; SAIC; CACI; the Carlyle Group; and others. Prior to our invasion of that country, conferences were held, showing companies how to belly up to the government trough. Propaganda was used to sell the American public on the war: WMDs; Saddam is a really bad man (that was true, but wasn't relevant to constitutional use of the military); the Iraqis want democracy (perhaps, but again, not relevant to constitutional use of the military); and, we have to go there to fight Al Qaeda, so we won't have to fight them here (I won't even comment on that one). Some Americans continue to believe that our leaders were sincere, but had bad intelligence reports. Perhaps, but I seriously doubt it. As mentioned previously, Saddam was about to accept

Euros for his oil; that's why he had to go. Plus, there's Brzezinski's *Grand Chessboard and Its Geostrategic Imperatives **for American Primacy*** [Emphasis added]. Our almost one billion dollar embassy in Iraq, the largest in the world, signifies that we

intended to be in that country for a long time. It's hard to say in 2022 whether or not we're gone for good. One rarely ever knows exactly where we continue to have special forces, or CIA Officers, and the like.

Corporate Globalization is taking place largely because of unelected corporatists and the use of propaganda. The world appears to be grinding slowly toward a world governance (or perhaps government) run by the Heads of mega corporations and other elites. Nevertheless, there does appear to be some hope for reversing the process. Not long ago in Bolivia, the municipal water system was privatized in a major city...due to a Structural Adjustment Program. The private contractor was Bechtel, a major transnational corporation. Within less than two months of taking over operation of the water system, Bechtel raised the consumer fees by fifty percent. The company also claimed rights to rain water and *all* the river water. The Bolivian people protested, to no avail. Finally, the People took to the streets en masse and eventually forced Bechtel out. [Under privatization in India, entire rivers have been sold to companies; the government then designates those living along a river and using the water as "water thieves".]

Many South American countries now are reevaluating their entire relationship with the WTO, the IMF, and the World Bank. All of the countries there that had bases, except for Columbia, have ousted the U.S. military. We have seven military bases in Columbia, ostensibly for the "War on Drugs"; however, it's more likely the bases are there to combat the country's 40-year-old insurgency, and to intimidate Maduro in Venezuela. [There was a successful coup against Chavez in 2002, but almost immediately, Venezuelans restored him to power.] Our Government appears to have various efforts to oust Maduro. I don't agree with what I know of his governing methods, but then, supposedly there's this thing called "sovereignty". Arguably, he is not a threat to the USA or the world in general.

When Kissinger was told that Allende was going his own way, Nixon's Secretary of State responded with, "Chile is a virus that might spread contagion...", meaning, Allende must be stopped before other Latin American countries get the idea that they, too, could be independent of the U.S. Consequently, the coup was carried out. Then began the testing of the neoliberal hypothesis.

One of my great concerns is that most of the Americans seem to be largely unaware of the implications of Corporate Globalization; and further, that they have almost no interest in discovering anything about it. Without identifying the problem and seeing how it came about, nothing effective can be done to solve it. Globalization (of the sort described here) is exporting manufacturing jobs and making "colonies" out of Third World countries. Furthermore, it is transferring wealth from our Middle Class to the Upper, Upper Class at an astonishing rate. Finally, it is creating a global economic and financial superstructure that is morphing into an <u>unelected</u> world government. It can be stopped if We the People wake up, get organized, and oppose the elites responsible. It appears to me that the first step in opposition is to oust the politicians who have approved the various "Free Trade Agreements" (such agreements are hardly free if they are *forced upon Third World countries*). We need to disengage from those agreements. Next, we should question loudly who it was that put the unelected World Trade Organization essentially in charge of all global trade, and expose their extortion of poor countries relative to privatization of public services. Voting with your wallet may very well have an impact on the transnational mega corporations, but it would help if it were a coordinated effort. Other boycott type activities surely are possible.

Corporate Globalization is a major aspect of neoliberal economics. The whole ball of wax is presented as some sort of perpetual motion machine which supposedly is superior to the biophysical world and unaffected by Laws of Nature. Then, too, its adherents claim their economic approach is lifting the world out of poverty and spreading

democracy. In reality, it's ruining the natural world, exploiting developing countries, and consigning the developed world to a monstrous rat race. On top of that, it's the cause of resource wars. When peaceful means fail, some countries go to war to secure access to oil, rare earth minerals, and soon, most likely fresh water.

This situation must change ASAP. That should be obvious to anyone with a pulse. I'd like to emphasize: none of this is some "vast conspiracy". Again, it's simply rich and powerful people with similar interests working mostly independently, but sometimes in concert with one another, toward a similar goal. For the most part, their actions are in plain sight. To them, it's simply "good business practice".

Finally, *prior to* the modern-day, mega business/neoliberal creation of a truly global market, it was known that such a market would create massive job disruption in much of the developed world. Shifting manufacturing to underdeveloped countries was almost certain to garner that result. To ameliorate the economic damage, the theorists envisioned the creation of more service or non-production employment in developed countries and *an expanded social safety net.* The latter did not happen; instead, austerity became the flavor of the day in many importing countries.

On top of that, job security vanished. Members of the emerging class of workers, the Precariat (see "Standing" in References), find themselves in part-time or temporary employment. Or both. Many have two or more jobs. Even those with full-time employment often have little-to-no job security, few benefits, and low wages/salaries. Trying to advance themselves, many times they are looking for a better job or retraining, and all the while getting deeper into the insecure rat race. Such are some of the perhaps unintended consequences of the neoliberal global market.

The proponents of neoliberal economics generally claim the economy is booming. It is booming for approximately the top 15-20% of income earners. It's really booming for the top one percent. Unfortunately, the bottom 80-85% barely are keeping up with price inflation. Most of them probably are doing that with overuse of credit cards.

Neoliberal proponents also have touted the fact that Globalization has resulted in less expensive goods being available now in developed countries. That's true. What they don't tell you is that those slippers or that coffee maker you bought for less money will break down about ten times faster than the one which was once manufactured domestically. Why? It's because the "globalized" goods are made with such cheap materials that they're essentially nothing but junk. On top of that, workers making such junk are paid very little for their labor, probably take little pride in their workmanship, and the result is not only junk, but junk with flaws as well. The cheaper prices are no bargain in the long run. Example: in 1958, I bought my first pair of new, domestically manufactured, Justin brand boots for $40. Except for being re-soled twice, those boots lasted thirty years. Three years ago, I bought a pair of new, made-in-China, Justin boots for $142. They are now falling apart in major ways. I've had similar experiences with coffee makers, slippers, shirts, bath towels, and several other consumer goods. I imagine other people have as well.

Appendix III: Politics

Politicians and their supporters generally adhere to one ideology or another. In some cases, they become almost fanatical "True Believers". That's a problem. Why? Mostly because they then essentially are closed-minded, dogmatic, and often "holier-than-thou". The problem really takes hold when a society is facing an existential threat, and one side of the political spectrum and/or the other won't budge in terms of dealing with the threat in an *effective* way. Our ongoing socio-eco-econ-ethical Crisis is the mother of all existential threats.

The Corporate State

Both FDR and Mussolini defined Fascism as: the marriage of the power of Big Business to the power of the State. An argument easily could be made for the following proposition: we've had a version of Soft Fascism in the USA for decades. The neoliberal policies of deregulation, privatization, austerity, suppression of free speech, the monopolization of the Tech Industry, etc. could not have been implemented without the tacit agreement of the Fed Government. The results have been social inequity, income inequality, ecological damage, financial crises, and an economic system which exacerbates all the above. And that system is still in operation.

In most cases, here's what happens regarding candidates for national political office in our Land. Such candidates are *selected, groomed, and financed* by the Super-Rich in mega corporations, especially by those in the Financial Sector of the economy. Once elected, the politicians rarely ever ignore those who helped get them into office. In such cases, that's often not beneficial to either Main Street or the natural ecosphere. In other words and plainly put, the political system is corrupt to the core.

Elections, the Oligarchy, and the Macro Level of Change

For decades and decades, "voting" in national elections in the USA has been largely or nearly pointless. Here's why. [NOTE – The 2020 election (and perhaps a few others) was an exception to the above contention. Trump was a straight-up, wannabe fascist, and had to go.]

1. Should some candidate other than an Oligarchy puppet manage to get elected, that person is fairly quickly marginalized, rendered ineffective, and/or smeared by the Political System. He/she is relegated to being little more than the butt of Establishment jokes and not taken seriously by the ruling elite.

2. Because of restrictive ballot-access laws created by Democrats & Republicans, any third Party rarely is able to get on the ballot in all 50 States. Even if an alternative Party does manage to do so, the Corporate Media Machine pretty much ignores them. In general, so do the major Parties.

3. The approximately 50% of eligible voters here who do actually vote in most national elections get the bulk of their information regarding candidates from political ads written by advertising agencies...in other words, from bald-faced propaganda. Issues are glossed over. Instead, the ad campaign mostly is about personalities.

4. Whoever runs the most ads usually wins. As a result, one election cycle at the national level can involve the expenditure of over a billion dollars. That's insanity, and is no way to select someone to represent We the People. Why? Because it fosters more & more reliance on money & propaganda...and results in more & more corruption.

5. The major political Parties essentially are two branches of the only political party of

any significance in America: the unnamed Mega Transnational Corporatist Party, or Mega Corporation Party. Yes, that's arguable; but as a keen observer of USA politics since **1956** (1956 is not a typo), that's how I see it.

6. No matter who gets elected at the national level, *Republican or Democrat*, they will promote, support, or at least not effectively oppose the following: institutionalized deficit spending; the Fed Reserve System of privately owned banks; the current, unconstitutional activities of the FBI relative to conducting searches without a proper warrant; other violations of the First and Fourth Amendments by the NSA, FBI, & CIA; protecting mega bank CEOs from criminal prosecution; the Fed Government's exemption of itself from International Law; so-called "Free Trade" Agreements; the greatest shift of wealth in history from the Middle Class to the upper one percent of adult Americans; unconstitutional wars; unconstitutional "Executive Orders"; the shipping of manufacturing overseas; corporate welfare; and on & on.

7. At the highest levels, Republicans-Democrats or NeoConservatives-Neoliberals do their very best to keep the general public from truly participating in government. Most all of them firmly believe in a Plutocracy, Oligarchy, or Meritocracy***. They are "True Believers" and convinced that unlike the "ignorant masses", they know best how to "run" the country. [***Note: Originally, the "Meritocracy" meant rule by *aristocrats* of merit.]

8. Once elected, a politician's first priority is to get re-elected. Bellying up to the Federal Trough is too sweet to risk losing. [That's one reason we desperately need Term Limits, something opposed by them all.]

In general, Plutocrats are committed to unlimited growth and overconsumption. If ever we are to reach sustainability goals on Spaceship Earth, we must stop allowing the Super-Rich to rule the roost. Step One would seem to be: get private *mega* donations out of political election campaigns.

The Plutocrats claim such money is "free speech". Not when it's given to candidates or their campaigns, it isn't. It's then nothing more than **corruption**, buying influence. The use of it would be free speech if the <u>donor</u> used it to run an ad in the media, expressing support for the candidate. This is not rocket science. It's merely common sense. Get mega money out of politics by peacefully "lobbying" any way you can imagine.

Not much will change if we keep doing nothing but "voting"; we've been doing so for well over 200 years. Things are only sliding downhill. "Voting" has accomplished very little for We the People...especially in the areas of jobs, wars, health care, and corruption. We're rapidly moving toward a NeoFeudal Society. The Working Poor is the fastest growing economic class in this country.

Before voting can make a significant difference, ballot-access laws must be changed. Campaign finance laws must be changed...or better yet, a Constitutional Amendment is needed relative to financing campaigns. None of this will happen without immense, peaceful pressure put on politicians; it is not in their best interests to change any of the current corruption.

Beginning steps toward a measure of sanity in USA politics:

1. stop drinking the Propaganda Kool Aid;

2. **peacefully and consist**ently insist that the Constitution be obeyed;

3. organize; and

4. peacefully resist, protest, and/or boycott.

[Imagine if at the next oligarchical election, only 2-10% voted. Now that is a message which could not be ignored. Unfortunately, it most likely won't happen...because of "True Believers" and Propaganda Kool Aid drinkers.]

5. Lobby for laws which remove private mega money from election campaigns.

6. When appropriate, discuss political corruption with people you know. Get people aware of the danger to democracy.

Politics in America - and probably most of the developed world - has been co-opted by mega transnational corporations. The Democratic and Republican Parties are almost totally irrelevant. Third Parties are as well. The system is rigged against them all. Until there are changes to ballot-access and campaign finance laws, that will continue. *In terms of formal politics, sorry to say, but the Oligarchy has won*...they have almost completely conquered us by Distraction and Division. That's only in terms of formal politics, though. There are other **peaceful** means to do battle. It pains me greatly to see people still stuck in formal politics, thinking they'll change things from the inside without a massive demand from Main Street. Sorry, but the probability is high it can't be done anymore; the Oligarchs/Plutocrats are nearly always in complete control of formal politics, and generally they love the status quo.

Many here were encouraged by the recent election of Joe Biden. While he's a gazillion times better than Trump, nevertheless he's a neoliberal, or at best, a puppet of neoliberals. How can I say that? Because I've studied his decades-long voting record. It overwhelmingly adheres to the principles of neoliberalism.

"Voting" is the opium of the masses in this country. Along with divisive propaganda and "bread & circuses", it has enabled Oligarchs to continue to rule this Land. Kicking Trump out of office was a good first step toward a bright future, but much more needs to be done. **We must transcend formal politics**. Significant change in the political system and our lives will come primarily from Movements originating *from the ground up,* not the top down. *Recent history has proven that to be true*. In addition, various Movements must join forces. Why? The reasons are simple: because everything is connected to everything else; and, solidarity – not fragmentation – is the key to success in bringing about positive change. Some of that joining together has started to take place. For example, some environmental groups at times have joined forces with groups of ranchers/farmers to defeat proposals by the oil industry which, if implemented, would have significantly damaged natural ecosystems. Another example: different groups have joined forces with Black Movements in order to foster policies favoring justice and social equity.

All big ideas and important changes started out with small steps. Then came organization, dedication, and persistence. If you really want to help organized human existence survive and thrive, then one of the things which must be done is to *change formal politics*. It best can be changed through the application of peaceful, persistent, and organized pressure on the Powers-That-Be. Main Street Movements are what bring about significant changes in the political system. By the way, that doesn't necessarily have to mean marching in the streets. Boycotts, local application of sustainability programs, Art, lobbying locally for the implementation of sustainability policies, advertising, writing, and more all can be effective tools for positive change. We must make a concerted effort before it's too late.

The Right and the Left

One of the main hindrances to effective political action in this country is the uncivil battle between the Right and the Left. The intolerance, hatred, and vitriol from the two factions often is beyond the pale. The "holier-than-thou" attitudes from each (at times) must change if we as a society ever are going to make *significant* progress toward genuine sustainability. More than ever, what the world needs now is *cooperation* rather than divisive competition.

The competitive battle often revolves around perceptions of "capitalism" versus "socialism". In my opinion, those perceptions too often are shallow, or worse, false. Consider the following.

The Left usually lumps the different types of capitalism all into one basket – plutocratic monopolies/egregious mega business/elitist trickle-down economics. The Right does something similar with socialism – it's all harsh totalitarianism to them. More often than not, it's considered communism. In both cases, the reality is much different.

Today's neoliberal capitalism is not Adam Smith's capitalism. Smith was for small business, not mega monopolies. Democratic socialism is quite different from dictatorial socialism. And here's the kicker in all of it: most governments (including ours in the USA) today have a **Mixed Economy** in their countries, part capitalism and part socialism.

When you use a publicly owned road, water system, library, museum, or public school, that's all due to socialism. The V.A. Health Care System is socialism, so too are Social Security and Medicare. If any two people engage in a private sale of private goods, that's capitalism. So, if I sell a painting of mine, or if this book results in even one sale, that's capitalism. When you go to a privately owned grocery store and purchase goods, that's due to capitalism.

The Right and the Left need to understand it isn't a matter of one or the other. **We have both, and we need both**. For example, if all roads were privately owned toll roads, that would be a nightmare. Some people could be refused use of them, or charges for use could go up for little to no reason. If all grocery stores were owned by the State, that too would be a nightmare. There would be no reason to have "sales" (offer bargains), or a wide variety of goods in order to attract more buyers. What for? Stores of any one type (e.g., grocery, hardware, clothing, etc.,) would all be more or less the same.

Bottom line: EVERYTHING being run by capitalists (private owners) simply isn't best for the good of all people. The same is true for everything being run by the State, i.e., socialism. **Economic systems have evolved into a mixture of both**. Unfortunately, at present that mixture is unbalanced in favor of mega business elites/plutocrats. Even worse, they are seriously pushing for private ownership of everything.

The political Right here (in the USA) was co-opted decades ago, first by neocons, then by neolibs... and most of the Right on Main Street appear to be completely unaware of the coup. They seem to think they're still in the "Conservative Movement". That movement, started by Russell Kirk in the 1950's, is long dead. The so-called conservatives of today are supporting an ideology which would be anathema to the original modern-day conservatives of the Kirk Era. That ideology they're supporting is corporatist-globalist-neoliberalism, and it doesn't want to conserve much of anything, except for maybe Plutocratic Rule. It's against Small Biz, anything rural, anything natural, and is working to utterly destroy Nature by favoring infinite growth at any cost to our finite habitat. Its goal is to have the entire world run by Big Biz Elites. Forget family farms, the mom-and-pop corner store, and any kind of level playing field in the business-financial-economic-labor arena. In general, neoliberals and the Super-Rich are against all of that. The kicker is this: **they finance most national political election campaigns.** That's why, in the USA, we've seen little progress toward sustainability over the past few decades.

Political prejudice reigns supreme (in general) among people <u>on the Right & the Left</u>, which means each views "the other side" as ALL (or at least, an overwhelming majority) having the same characteristics as the few who are running the show on each side. As long as that continues, we will not have a civil society. The truth is, no group ever has been as homogeneous as what the Right and Left say the other side is today.

I've seen ranchers on the Right and environmentalists on the Left work together to oppose mega corporate/government projects which would have harmed the commons. They had to overcome deep prejudices about each other, and they did. Unfortunately, such collaboration is extremely rare because of those deep prejudices.

We all should make sincere efforts to overcome political stereotypes regarding people who do not share our particular beliefs. There are some very good people all across the political spectrum. Why focus only on those who seemingly are filled with hate? I

believe they are an extreme minority. It really is time for civil political discourse, and for efforts to find some common ground.

Here's the main thing wrong with the general political Left in the social, educational, & political spheres of life--- https://www.youtube.com/watch?v=snqXOvnHzcQ ["The Stunning Fragility & Vindictive Political Correctness of Today's Students - Jonathan Haidt"] and, although it's well-intentioned, it's creating a fragility in our youth that's beyond the pale. Plus, it's accompanied by an almost virulent self-righteousness. The speakers here expose it, & explain why it's damaging...not only to our youth, but also to the country in general. The clip is only 13.5 minutes in length.

The subject matter at the above link includes: egregious, over-the-top, political correctness; the coddling of kids; gross intolerance of views other than those promulgated by the Left; "slurring" rather than debating in a reasonable manner; and a subtle **self-righteousness** regarding social and political issues. None of that bodes well for a transition to sustainability.

By the way, the political Right (in general) is as bad as, or worse than, the Left regarding this whole shebang... at least, most of it. *The result is that the fracturing of our society gets worse & worse every year.* Civility is well on its way to being totally gone. It's being replaced by not only rudeness, but intolerance and **violence** as well. The Corporate Media seem to focus on such activities, but rarely (if ever) ask "Why is this happening?". Undoubtedly, the answer contains many factors, complex factors; nevertheless, we need to figure this thing out. If we don't, I believe our society will continue to break down, perhaps even to the point of Martial Law being imposed. It costs little to nothing to seriously and *rationally* consider an opposing point-of-view.

A starting point is this: **every side in any significant disagreement passionately believes that its side is the one "in the right", or "best for most people".** The opposing sides need to find out why "the other side" thinks that way... rather than simply slurring them and giving them sound-bite retorts.

Politicians, institutions of all kinds (religious, educational, social, Law enforcement, non-profit, for-profit, etc.), and all of us "common" people need to wake up and re-double our efforts to promote reasonable & civil debate, tolerance, and respect for individuals with opposing views. I've seen decades of decay in that regard. It doesn't bode well for the future of our Land.

So, though it may be extremely difficult to do, I propose to all "True Believers" to (at least once in awhile) step outside the "For or Against" paradigm and make a herculean effort to consider issues without having any prejudgments. I like Socrates' observation on Wisdom: "The only true wisdom is in knowing you know nothing.". Let's be more tolerant of each other's views...and start learning. It costs little-to-nothing to do so. What we shall reap is civility, knowledge of common ground, and perhaps a much, much better world for us, our children, and our grandchildren.

References

Almond, R.E.A., Grooten, M., and Petersen, T. (Editors). 2020. *Living Planet Report 2020 – Bending the curve of biodiversity loss,* World Wildlife Fund (WWF). pp. 6-35, 50-73, 112-135.

Bacevich, Andrew. 2020. *The Age of Illusions: How America Squandered Its Cold War Victory,* Metropolitan Books.

_____. 2005. *The New American Militarism – How Americans Are Seduced by War,* Oxford University Press.

_____. April 18, 2021. *Op-Ed: We don't need a new Cold War with China*, Los Angeles Times.

Bardgett, Richard D. and Wardle, David A. 2003. *HERBIVORE-MEDIATED LINKAGES BETWEEN ABOVEGROUND AND BELOWGROUND COMMUNITIES*, Ecology, Volume 84, Number 9.

Barnes, Peter. 2001 and 2003. *Who Owns the Sky? - Our Common Assets and the Future of Capitalism*, Island Press.

_____. 2006. *Capitalism 3.0: a guide to reclaiming the commons*, Berrett-Koehler Publishing. pp. 65-116.

Bello, J. et al. September 2020. *Providing an enabling environment to promote the Sustainable Development Goals: Coventry University's experience*, Emerald Open Research.

Belmonte-Urena, Luis Jesus. et al. July 2021. *Circular economy, degrowth and green growth as pathways for research on sustainable development goals: A global analysis and future agenda*, Ecological Economics, Volume 185.

Bernays, Edward. 1928. *Propaganda – Molding the Public Mind,* Horace Liveright Publisher.

Bernier, Austen K. 2016. *Neoliberalism and the Environmental Movement: Contemporary Considerations for the Counter Hegemonic Struggle*, CU Scholar, Undergraduate Honors Theses, Paper 1013. pp. 31-81, 86-92.

Biggs, Eloise M. et al. December 2015. *Sustainable development and the water–energy–food nexus: A perspective on livelihoods,* Environmental Science & Policy, Volume 54. pp. 389-397.

Bindi, Marco. et al. 2018. *Impacts of 1.5 Degrees Celsius of Global Warming on Natural and Human* Systems, IPCC Special Report, Chapter 3. [Intergovernmental Panel on Climate Change]

Blanco, Eduardo. et al. January 2021. *Urban Ecosystem-Level Biomimicry and Regenerative Design: Linking Ecosystem Functioning and Urban Built Environments*, Sustainability, Volume 13, Issue 1.

Bontz, Scott (Edited by). Fall, 2021. *Land Report – Number 131*, The Land Institute. pp. 4-28.

Bourban, Michel. Online First: January 18, 2022. *Strong Sustainability Ethics*, Environmental Ethics. [Soon to be in the Journal as hard copy]

Boyce, James K. November 28, 2020. *The Case for Universal Property*, Scientific American.

Bradshaw, Corey J. A. Ehrlich, Paul R. et al. January 2021. *Underestimating the Challenges of Avoiding a Ghastly Future*, Frontiers in Conservation Science. Perspective.

Bretschger, Lucas. October 2021. *Getting the Costs of Environmental Protection Right: Why Climate Policy is Inexpensive in the End*, Ecological Economics, Volume 188.

Brown, Charles S. Winter 1995. *Anthropocentrism and Ecocentrism: the Quest for a New Worldview*, The Midwest Quarterly, Volume 36, Number 2.

Brown, Lester R. 2008. *Plan B 3.0 – Mobilizing to Save Civilization*, W.W. Norton & Company. pp. 133-139, 153-163, 176-182, 194-204, 239-258, 267-272, 280-287.

Brown, L. R. Flavin, C. French, H. et al. 1999. *State of the World, 1999, A Worldwatch Institute Report on Progress Toward a Sustainable Society*, W.W. Norton & Company. pp. 41-59, 78-95, 133-150, 169-188.

Buscher, Bram. April 2021. *Planning for a world beyond COVID-19: Five pillars for post-neoliberal development*, World Development, Volume 140.

Cairns, John Jr. 2004. *Eco-Ethics and Sustainability Ethics*, Inter-Research (Publisher). pp. 189-202, 220-252, 274-277, 288-306.

Campanale, Claudia. et al. February 2020. *A Detailed Review Study on Potential Effects of Microplastics and Additives of Concern on Human Health*, International Journal of Environmental Research and Public Health, Volume 17, Number 4.

Carson, Rachel. 1962. *Silent Spring*, Houghton Mifflin. pp. 15-52, 85-102, 154-172, 187-198, 245-261, 277-300.

Castree, Noel et al. 2009. *A Companion to Environmental Geography,* Blackwell Publishing, Ltd.

Clarke, Felix. November 2021. Book Review: *Entropy Law, Sustainability, and Third Industrial Revolution, R. Sengupta, Oxford University Press, Oxford, UK (4 August 2020)*, Ecological Economics, Volume 189.

Clayton, P. and W. A. Schwartz. 2019. *What is Ecological Civilization? - Crisis, Hope, and the Future of the Planet*, Process Century Press. pp. 53-68, 82-93, 114-143.

Collins, Chuck. 2016. *Born on Third Base – A One Percenter Makes the Case for Tackling Inequality, Bringing Wealth Home, and Committing to the Common Good,* Chelsea Green Publishing. pp. 59-100, 141-203.

Commoner, Barry. 1971. *The Closing Circle – Nature, Man, and Technology,* Knopf Press and Bantam Books. pp. 14-48, 112-124, 140-215, 250-292.

_____. 1976. *The Poverty of Power – Energy and the Economic Crisis,* Alfred A. Knopf (Publisher). pp. 6-29, 76-112, 221-249.

Corsa, Andrew J. 2021. *John Cage, Henry David Thoreau, Wild Nature, Humility, and Music,* Environmental Ethics, Volume 43, Number 3.

Costanza, Robert, et al. 1997. *An Introduction to Ecological Economics,* St. Lucie Press (a publishing imprint of CRC Press). pp. 80-100, 108-139, 206-216.

Curry, Patrick. 2011. *Ecological Ethics – An Introduction,* Second Edition, Polity Press. pp. 28-49, 184-243, 253-266.

Czech, Brian. 2004. *A Chronological Frame of Reference for Ecological Integrity and Natural Conditions*, Natural Resources Journal, Volume 44, Number 4.

_____ (Edited by). 2020. *Best of the Daly News – Selected Essays from the Leading Blog in Steady State Economics, 2010-2018,* Steady State Press. pp. 3-56, 65-66, 82-84, 93-95, 107-110.

_____. 2021. *Supply Shock – Economic Growth at the Crossroads and the Steady State Solution*, Steady State Press (an imprint of the Center for the Advancement of the Steady State Economy – CASSE). pp. 51-116, 137-169, 275-327.

_____. January 13, 2022. *True Conservation: A 21st Century Vision for the Next Director of the U.S. Fish and Wildlife Service*, Steady State Herald, SteadyState.org

Dallek, Robert. 2017. *Franklin D. Roosevelt: A Political Life,* Viking (an imprint of Penguin Random House).

Daly, Herman E. 1996. *Beyond Growth: The Economics of Sustainable Development,* Beacon Press.

_____. 1999. *Ecological Economics and the Ecology of Economics,* Edward Elgar Publishing. pp. 27-73, 77-105, 123-131.

_____ and Farley, Joshua. 2004. *Ecological Economics, Principles and Applications,* Island Press. pp. 19-35, 64-76, 121-122, 201-220, 228-244, 323-342, 365-372.

D'Amato, D. and Korhonen, J. October 2021. *Integrating the green economy, circular economy and bioeconomy in a strategic sustainability framework*, Ecological Economics, Volume 188.

Dhara, Chirag, and Singh, Vandana. June 2021. *The Delusion of Infinite Economic Growth,* Scientific American.

Duff, Hannah and Hegedus, Paul B. et al. 2022. *Precision Agroecology*, <u>Sustainability</u>, Volume 14, Number 1.

Ehrlich, Paul R. December 2014. *Human Impact: the ethics of I=PAT,* Stanford University, Center for Conservation Biology.

_____. February 2008. *Key issues for attention from ecological economists,* <u>Environment and Development Economics</u>, Volume 13, Number 1.

_____. 1997. *A World of Wounds: Ecologists and the Human Dilemma,* Ecology Institute. pp. 51-93, 113-151.

_____. March 2020. *Paul R. Ehrlich: A pandemic, planetary reckoning, and a path forward,* <u>Environmental Health News</u>. [Online]

Epstein, Graham et al. June 2015. *Institutional fit and the sustainability of social–ecological systems*, <u>Current Opinion in Environmental Sustainability</u>, Volume 14, pp. 34-40.

Fanning, Andrew L. et al. January 2022. *The social shortfall and ecological overshoot of nations*, <u>Nature Sustainability</u>, Volume 5.

Faria, Luiz R. R. July 2020 and April 2021. *Conservation optimism and reckoning with the future*, <u>Conservation Biology</u>. pp. 745-747.

Farrar, John et al. 2003. *HOW ROOTS CONTROL THE FLUX OF CARBON TO THE RHIZOSPHERE*, <u>Ecology</u> , Volume 84, Number 4.

Fazey, Ioan et al. June 2018. *Ten essentials for action-oriented and second order energy transitions, transformations and climate change research,* <u>Energy Research & Social Science</u> Volume 40, pp. 54-70.

Fernandez, Emilio et al. March 2019. *Operational principles of circular economy for sustainable development: Linking theory and practice*, <u>Journal of Cleaner Production</u>, Volume 214.

Fioramonti, Lorenzo. et al. February 2022. *Wellbeing economy: An effective paradigm to mainstream post-growth policies?*, <u>Ecological Economics</u>, Volume 192.

Fuller, R. Buckminster. 1969 (New Edition, 2020). *Operating Manual for Spaceship Earth,* Lars Muller Publishers. pp. 57-106, 121-138.

Ganter, Carl. January 16, 2015. *Water Crises Are a Top Global Risk*, World Economic Forum. [Online]

Georgescu-Roegen, Nicholas. 1971. *The Entropy Law and the Economic Process,* Harvard University Press.

Gill, Joel C. et al. June 2019. *The role of Earth and environmental science in addressing sustainable development priorities in Eastern Africa*, <u>Environmental Development</u>, Volume 30. pp. 3-20.

Gong, Hongjing et al. December 16, 2021. *Emerging Global Ocean Deoxygenation Across the 21st Century*, <u>Geophysical Research Letters</u>, Volume 48, Issue 23.

Goudie, Andrew S. 2019. *Human Impact on the Natural Environment – Past, Present and Future – Eighth Edition*, John Wiley & Sons. pp. 267-281, 341-359.

Grossa, John and Marsh, William M. 2001. *Environmental Geography: Science, Land Use and Earth Systems,* John Wiley & Sons, Inc. pp. 44-48, 81-102, 134-155, 162-182, 199-206, 227-234, 264-281, 336-355.

Gulyas, Boglarka and Edmonson, Jill. January 2021. *Increasing City Resilience through Urban Agriculture: Challenges and Solutions in the Global North,* <u>Sustainability</u>, Volume 13, Issue 3.

Hachaichi, Mohamed and Baouni, Tahar. February 2020. *Downscaling the planetary boundaries (Pbs) framework to city scale-level: De-risking MENA region's environment future* [MENA = Middle East & North Africa], <u>Environmental and Sustainability Indicators</u>, Volume 5.

Hammond, Debora. 2010. *The Science of Synthesis: Exploring the Social Implications of General Systems Theory*, University Press of Colorado.

Hanna, Elie and Comin, Francisco. October 2021. *Urban Green Infrastructure and Sustainable Development: A Review*, Sustainability, Volume 13, Issue 20.

Hawken, Paul. et al. 1999. *Natural Capitalism – Creating the Next Industrial Revolution,* Little, Brown and Company. pp. 48-61, 82-110, 190-212, 285-308.

Heinberg, Richard. 2011. *The End of Growth – Adapting to Our New Economic Reality,* New Society Publishers. pp. 6-20, 56-69, 105-153, 241-254, 268-283.

Herrfahrdt-Pahle, Elke. et al. July 2020. *Sustainability transformations: socio-political shocks as opportunities for governance transitions,* Global Environmental Change, Volume 63.

Hickey, Colin and Robeyns, Ingrid. December 2020. *Planetary justice: What can we learn from ethics and political philosophy?*, Earth System Governance, Volume 6.

Jensen, Robert. 2020. *Who is we?*, The Ecological Citizen, Volume 4, pp. 57-61.

Johnson, Justin A. et al. September 2021. *Energy matters: Mitigating the impacts of future land expansion will require managing energy and extractive footprints*, Ecological Economics, Volume 187.

Karlsson, Mikael, and Edvardsson Bjornberg, Karin. August 2020 and April 2021. *Ethics and biodiversity offsetting,* Conservation Biology. pp. 578-586.

Kates, Robert W. February 2016. *Sustainability Science*, International Encyclopedia of Geography: People, the Earth, Environment and Technology, Wiley Online Library, John Wiley & Sons.

Keller, David R. (Edited by). 2010. *Environmental Ethics – The Big Questions,* Wiley-Blackwell (A John Wiley & Sons Publication). pp. 77-81, 98-109, 137-142, 193-210, 268-274, 352-358, 380-411, 426-446, 509-533.

Kennedy, Christopher. February 2022. *The intersection of Biophysical Economics and Political Economy*, Ecological Economics, Volume 192.

Kinne, O. 2002. *Eco-ethics further developed text: 01.05.2002*. EEIU Brochure, Inter-Research, Oldendorf/Luhe.

Kolbert, Elizabeth. 2006. *Field Notes From a Catastrophe – Man, Nature,and Climate Change*, Bloomsbury USA. pp. 136-144, 183-199.

Leach, Melissa. et al. November 2018. *Equity and sustainability in the Anthropocene: a social–ecological systems perspective on their intertwined futures*, Global Sustainability, Cambridge University Press.

Lehmann, Paul. et al. July 2021. *Managing spatial sustainability trade-offs: The case of wind power*, Ecological Economics, Volume 185.

Lent, Jeremy. 2021. *The Web of Meaning – Integrating Science and Traditional Wisdom to Find Our Place in the Universe*, New Society Publishers. pp. 121-148, 260-289, 349-382.

Leopold, Aldo. 1949 (Ballantine Edition, 1966). *A Sand County Almanac,* Oxford University Press and Ballantine Books. pp. 237-295.

Limburg, Karin and Costanza, Robert (Editors). 2010. *Ecological Economics Reviews,* ANNALS of the New York Academy of Sciences, Volume 1185. Published by Blackwell Publishing on behalf of the NY Academy. pp. 11-26, 54-96, 119-131, 164-193, 225-244.

Lunstrum, Elizabeth and Bose, Pablo S. 2022. *Environmental Displacement in the Anthropocene*, Annals of the American Association of Geographers, Pre-print online version.

Marsh, William M. and Grossa, Jr., John. 2002. *Environmental Geography – Science, Land Use, and Earth Systems*, Second Edition, John Wiley & Sons, Inc.

Marselle, Melissa. et al. May 2021. *Pathways linking biodiversity to human health: A conceptual framework*, Environment International, Volume 150.

Matson, Noah P. 2004. *Maintaining the Biological Integrity, Diversity, and Environmental Health of the National Wildlife Refuge System*, Natural Resources Journal, Volume 44, Number 4.

Mattis, Kristine. August 2018. *Dying of Consumption While Guzzling Snake Oil: The Environment Crisis Requires Overhauling Our Corporate Industrial Civilization*, CommonDreams.org

_____. December 2018. *An Economy That Does Not Consider Ecology Is Not Sustainable*, CommonDreams.org

Maynard, Robin. 2021. *Overpopulation denial syndrome*, The Ecological Citizen, Volume 5, Number 1. pp. 23-28.

McKibben, Bill. 2007. *Deep Economy – The Wealth of Communities and the Durable Future*, Henry Holt and Company. pp. 46-66, 95-108, 129-158, 177-226.

_____. 1989. *The End of Nature,* Random House, and Anchor Books (Doubleday). pp. 47-70, 95-117, 139-149, 171-217.

Melgar-Melgar, Rigo E. and Hall, Charles A.S. March 2020. *Why ecological economics needs to return to its roots: The biophysical foundation of socio-economic systems*, Ecological Economics, Volume 169.

Miller, Todd. et al. October 2021. *Global Climate Wall – How the world's wealthiest nations prioritise borders over climate action*, Transnational Institute, Amsterdam. pp. 1-65.

Molles, Manuel and Sher, Anna. 2019. *Ecology: Concepts and Applications,* 8th Edition, McGraw Hill Publishing.

Murphy, Thomas W., Jr. 2021. *Energy and Human Ambitions on a Finite Planet - Assessing and Adapting to Planetary Limits*, eScholarship University of California.

Okushima, Shinichiro. September 2021. *Energy poor need more energy, but do they need more carbon? Evaluation of people's basic carbon needs*, Ecological Economics, Volume 187.

Orr, David W. 1994. *Earth in Mind – On Education, Environment, and the Human Prospect,* Island Press. pp. 35-39, 60-63, 74-77, 104-111, 131-152, 154-170.

Pal, R. et al. 2006. *Degradation and effects of pesticides on soil microbiological parame ters – a review,* <u>International Journal of Agricultural Research,</u>

Volume 1, pp. 240-258.

Poitras, Geoffrey. June 2021. *Rhetoric, epistemology and climate change economics,* <u>Ecological Economics,</u> Volume 184.

Prashad, Vijay. September 2020. *Washington Bullets: A History of the CIA, Coups, and Assassinations,* Monthly Review Press.

Preston, Jr., Richard J. 1970. *North American Trees,* The M.I.T. Press. pp. ix-x, xiii-xxiii.

Quigley, Carroll. May 1968. *Needed: A Revolution in Thinking,* <u>National Education Association Journal,</u> Volume 57. pp. 8-10.

Rabbi, M. F. et al. 2021. *Food Security and Transition Towards Sustainability,* <u>Sustainability,</u> Volume 13, Number 22.

Ratcliffe, Jennie M. 2021. *Nothing Lowly in the Universe,* The Crundale Press. pp. 157-176, 195-213, 245-275.

Rees, William E. October 1992. *Ecological Footprints and appropriated carrying capacity: what urban economics leaves out,* <u>Environment and Urbanization,</u> Volume 4, Number 2.

_____ and Wackernagel, Mathis. 1996. *Our Ecological Footprint: Reducing Human Impact on the Earth,* New Society Publishers. pp. 7-29, 40-57, 125-140, 149-157.

_____. June, 2015. *Economics vs. the Economy,* Great Transition Initiative [Publisher, online version].

Ripple, William J. et al. September 2021. *World Scientists' Warning of a Climate Emergency 2021,* <u>BioScience,</u> Volume 71, Issue 9.

Rissing, Andrea. 2021. *"We feed the world": the political ecology of the Corn Belt's driving narrative*, <u>Journal of Political Ecology</u>, Volume 28. pp. 472-487.

Rodeiro, Manuel. 2021. *Justice and Ecocide: A Rawlsian Account*, <u>Environmental Ethics</u>, Volume 43, Number 3.

Ruault, Jean-Francois et al. January 2022. *A biodiversity-employment framework to protect biodiversity*, <u>Ecological Economics</u>, Volume 191.

Sareen, Siddarth and Wolf, Steven A. July 2021. *Accountability and sustainability transitions*, <u>Ecological Economics</u>, Volume 185.

_____, et al. 2022. *Multiscalar Practices of Fossil Fuel Displacement*, <u>Annals of the American Association of Geographers</u>. Pre-print online version.

Sarkodie, Samuel A. January 10, 2021. *Environmental performance, biocapacity, carbon & ecological footprint of nations: Drivers, trends and mitigation options*, <u>Science of The Total Environment</u>, Volume 751.

Schumacher, E. F. 1973. *Small is Beautiful – Economics as if People Mattered,* Harper & Row Publishers. pp. 53-62, 102-117, 134-145.

Slater, Thomas. et al. January 2021. *Review article: Earth's ice imbalance*, <u>The Cryosphere</u>, Volume 15. European Geosciences Union. pp. 233-246.

Smith, Robert Leo. 1996. *Ecology And Field Biology,* Fifth Edition, HarperCollins College Publishers. pp. 130-142, 151-194, 361-409, 633-654.

Sparenberg, David. 2021. *Earth Spirit – Confronting the Crisis – Essays and Meditations on Eco-Spirituality*, John Hunt Publishing. pp. 9-11, 18-20, 32-35, 83-84.

Standing, Guy. June 25, 2021. *Basic Income Pilots: Uses, Limitations and Design Principles*, <u>Basic Income Studies</u>, Volume 16, Issue 1.

_____. November 20, 2014. *The Precariat*, <u>Contexts</u>, Volume 13, Issue 4.

Steffen, Alex (Edited by). 2006. *Worldchanging – A User's Guide for the 21st Century,* Harry N. Abrams, Inc. pp. 307-315, 337-341, 481-488, 514-536.

Suzuki, David and McConnell, Amanda. 1999. *The Sacred Balance –*

Rediscovering Our Place in Nature, Greystone Books – The Douglas & Mcintyre Publishing Group. pp. 76-104, 184-200, 207-240.

Taibbi, Matt. 2010. *Griftopia: Bubble Machines, Vampire Squids, and the Long Con That Is Breaking America,* Spiegel & Grau Publishers (a former publishing imprint of Penguin Random House).

Trantas, Nikos. 2021. *Could "degrowth" have the same fate as "sustainable development"? A discussion on passive revolution in the Anthropocene age*, Journal of Political Ecology, Volume 28. pp. 225-245.

Tolle, Eckhart. 2005. *A New Earth – Awakening to Your Life's Purpose*, Penguin Books/Plume Books. pp. 8-23, 25-84, 279-309.

Tschiderer, Luca. July 2021. Book Review - *The Birth of Energy: Fossil Fuels, Thermodynamics, and the Politics of Work, Cara New Daggett, Duke University Press, Durham NC (2019)*, Ecological Economics, Volume 185.

Vassallo, Paulo. et al. April 2021. *Biophysical Accounting of Forests' Value Under Different Management Regimes: Conservation vs. Exploitation*, Sustainability, Volume 13, Issue 9.

Villamayor-Tomas, Sergio. et al. April 2022. *Social Movements and Commons: In Theory and in Practice*, Ecological Economics, Volume 194.

Visalli, Dana. March 2021. *A Climate of Change,* MAHB Blog (Millennium Alliance for Humanity and the Biosphere) at Stanford University.

Wackernagel, M. et al. July 2017. *Making the Sustainable Development Goals Consistent with Sustainability*, Frontiers in Energy Research.

Washington, Haydn (Edited by). 2020. *Ecological Economics – Solutions for the Future,* Published by the Editor. pp. 25-90, 188-218, 309-356. [Book arose from the 2019 Australia New Zealand Society for Ecological Economics Conference held at RMIT University, Melbourne Australia in November, 2019.]

_____ and Maloney, Michelle. March 2020. *The need for ecological ethics in a new ecological economics*, Ecological Economics, Volume 169.

Watts, Alan. 1971. *Does It Matter? – Essays on Man's Relation to Materiality,* New World Library. pp. 1-53, 99-128.

Werner, Marion. et al. 2022. *The Glyphosate Assemblage: Herbicides, Uneven Development, and Chemical Geographies of Ubiquity*, <u>Annals of the American Association of Geographers</u>, Volume 112, Number 1.

Weston, Burns H. and Bollier, David. 2013. *Green Governance – Ecological Survival, Human Rights, and the Law of the Commons,* Cambridge University Press. pp. 50-76, 112-120, 123-154, 226-256.

Wilson, Edward O. 2016. *Half-Earth: Our Planet's Fight for Life,* Liveright Publishing, a division of W. W. Norton & Company.

_____. 2002. *The Future of Life,* Alfred A. Knopf (Publisher). pp. 22-41, 79-102, 149-189.

Wolff, Richard D. November 5, 2021. *Why the Troubled U.S. Empire Could Quickly Fall Apart*, <u>BRAVE NEW EUROPE NEWSLETTER</u>. [ONLINE]

www.ingramcontent.com/pod-product-compliance
Lightning Source LLC
Chambersburg PA
CBHW081125170526
45165CB00008B/2549